ENCHANTED OBJECTS

DESIGN, HUMAN DESIRE,
AND THE INTERNET OF THINGS

DAVID ROSE

SCRIBNER

NEW YORK   LONDON   TORONTO   SYDNEY   NEW DELHI

Scribner
A Division of Simon & Schuster, Inc.
1230 Avenue of the Americas
New York, NY 10020

First Scribner hardcover edition July 2014

SCRIBNER and design are registered trademarks of The Gale Group, Inc.,
used under license by Simon & Schuster, Inc., the publisher of this work.

For information about special discounts for bulk purchases,
please contact Simon & Schuster Special Sales at 1-866-506-1949
or business@simonandschuster.com.

The Simon & Schuster Speakers Bureau can bring authors to your live event.
For more information or to book an event contact the Simon & Schuster Speakers
Bureau at 1-866-248-3049 or visit our website at www.simonspeakers.com.

Interior design by Nathan Hass
Jacket design by Tal Goretsky and Kyle Poff
Jacket illustration by Brian Levy

Manufactured in the United States of America

1   3   5   7   9   10   8   6   4   2

ISBN 978-1-4767-2563-5
ISBN 978-1-4767-2565-9 (ebook)

Photograph and illustration credits appear on page 303.

To my children
Rayna and Adam,
and to yours.

The vision for a more humane interface
between technology and people
is in your hands.

"Any sufficiently advanced technology is indistinguishable from magic."

—Arthur C. Clarke

"Study the past, if you would divine the future."

—Confucius

CONTENTS

CONTENTS

INTRODUCTION

WHAT MAKES SOMETHING magical? Enchanted?

I'm not talking about deceptive magic—tricks and sleight of hand. This book is about how to strategically design and develop products that are engaging and essential, that resonate with the latent needs of those who use them, and that create an emotional connection with us human beings. I have spent nearly twenty years developing Internet-connected things (toys, furniture, lighting fixtures, jewelry, and more), and I remain disappointed that so few products succeed in enchanting us. Instead, they are difficult to understand, frustrating to use, overwrought with features. They diminish rather than empower us.

This book is meant to catalyze the imagination of designers, business strategists, and technologists to craft more delightful products and more enchanted experiences—and to remind everyone who uses Internet-connected things (which is all of us) that we should expect more from the tools, devices, and playthings that are such an enormous part of our lives.

What's the secret to creating technology that is attuned to the needs and wants of humans? The answer can be found in the popular stories and characters we absorb in childhood and that run through our cultural bloodstream: Greek myths, romantic folktales, comic book heroes, Tolkien's wizards and elves, Harry Potter's entourage, Disney's sorcerers, James Bond, and Dr. Evil. They all employ enchanted tools and objects that help them fulfill fundamental human drives. In this book, I link the fictions and fantasies that so beautifully express these desires and the role of modern inventions. My goal is to change the way you think about computers and computer-driven things and how we interact with them.

I teach at MIT's Media Lab, where one of the great benefits of my work is the constant stream of visitors who pass through day after day: business executives, dignitaries, musicians, architects, designers, technologists, and the occasional Hollywood producer. They come in search of insight into how our lives might be different in the future and how technological change might affect their work.

One spring afternoon, J. J. Abrams—producer of the television series *Lost, Fringe,* and two *Star Trek* movies—stopped by to see demonstrations of prototype technologies and to talk about magic and science fiction. A few days after his visit he sent an email in which he asked a provocative question: "Fifty years from now, what will computers be called?"

He got plenty of responses from my students and colleagues. Syn. Neuro. Heisenberg. Mother. Your Excellence. One student, Katherine, replied, "I think they will be called nothing. They will 'be' us and power everything under the sun." And César agreed: "Probably we will just say something like 'I'm going in,'" and people will understand what they mean.

The conversation that Abrams spurred was not really about names but rather about the relationship we will have—and want to have—with future technology. Do we want more tablets and screens? How do we feel about robots and wearables? What about enchanted everyday objects?

What personality do we want our technologies to possess? Domineering or polite? Should our technologies look cold or cute? Do we want to interact with them as smart tools or as caring agents? Should every child be required to learn to code or is a zero learning curve the ideal? Do we want computers to become more human or humans to become more like computers?

I hope to shed light on these issues through the stories of some forty Internet-connected things and to explore the ramifications of how the human-machine interface impacts the design of wearable technology, medical devices, vehicles, communication tools, musical instruments, drawing instruments, our homes, our workplaces, and, in the future, almost every nonhuman element of our lives.

In Part I of the book, I describe the four likely technological futures: Terminal World, prosthetics, animism, and enchanted objects. In Part II, I explore the six human drives—omniscience, telepathy, safekeeping, immortality, teleportation, and expression—and the dialectic interplay of the fictions and inventions associated with those drives. Part III is about how to design enchantment, including how to think about the major "abilities" of enchanted objects—including gestureability and glanceability—and how to approach the design process as a "ladder" of enchantment, from augmentation to story-ification. In Part IV, I look ahead at how larger systems—our homes, our workplaces, and our cities—might be transformed through enchanted objects. I leave readers with six fantasies of what I would love to see come next.

While this book is meant to appeal to both general readers as well as specialists, I'm particularly interested in your willingness to flex and consider the world from three perspectives: technology, design, and business. It takes a polyglot to understand and make smart decisions about human-centered products, so your ability to understand and communicate with other scientists, engineers, designers, psychologists, executives, and entrepreneurs—as well as customers and users—is essential to taking part in the next wave of the Internet.

Welcome to the age of enchanted objects.

ENCHANTED OBJECTS

MY NIGHTMARE

I HAVE A recurring nightmare. It is years into the future. All the wonderful everyday objects we once treasured have disappeared, gobbled up by an unstoppable interface: a slim slab of black glass. Books, calculators, clocks, compasses, maps, musical instruments, pencils, and paintbrushes, all are gone. The artifacts, tools, toys, and appliances we love and rely on today have converged into this slice of shiny glass, its face filled with tiny, inscrutable icons that now define and control our lives. In my nightmare the landscape beyond the slab is barren. Desks are decluttered and paperless. Pens are nowhere to be found. We no longer carry wallets or keys or wear watches. Heirloom objects have been digitized and then atomized. Framed photos, sports trophies, lovely cameras with leather straps, creased maps, spinning globes and compasses, even binoculars and books—the signifiers of our past and triggers of our memory—have been consumed by the cold glass interface and blinking search field. Future life looks like a *Dwell* magazine photo shoot. Rectilinear spaces, devoid of people. No furniture. No objects. Just hard, intersecting planes—Corbusier's Utopia. The lack of objects has had an icy effect on us. Human relationships, too, have become more transactional, sharply punctuated, thin and curt. Less nostalgic. Fewer objects exist to trigger storytelling—no old photo albums or clumsy watercolors made while traveling someplace in the Caribbean.

In my nightmare, the cold, black slab has re-architected everything—our living and working spaces, our schools, airports, even bars and restaurants. We interact with screens 90 percent of our waking hours. The result is a colder, more isolated, less humane world. Perhaps it is more efficient, but we are less happy.

Marc Andreessen, the inventor of the Netscape browser, said, "Software is eating the world." Smartphones are the pixelated plates where software dines.

Often when I awake from this nightmare, I think of my grandfather Otto and know the future doesn't have to be dominated by the slab. Grandfather was a meticulous architect and woodworker. His basement workshop had many more tools than a typical iPad has apps. He owned power tools: table saw, lathe, band saw, drill press, belt sander, circular sander, jigsaw, router. And hand tools: hundreds of hammers, screwdrivers, wrenches, pliers, chisels, planes, files, rasps. Clamps hung from every rafter. Strewn around his architectural drawings were T squares, transparent triangles, hundreds of pencils and pens, stencils for complex curves, compasses, and protractors of every size.

The diversity of wood-working tools in my grandfather's workshop, or utensils in a kitchen, or shoes in your closet, prove our presence for specialization. This debunks the myth of technology convergence.

I don't recall my grandfather ever complaining about having too many tools. Or dreaming of tool convergence—wishing some singular mother-of-all-tools would come along to replace them. Redundancy abounded. Specialization was prized. When carving, he would lay out a line of chisels that, to my untrained eye, looked pretty much the same.

He would switch rapidly from tool to tool, this one for a smaller-radius cut, this one to take out more material, this one for a V-shaped cut. As a five-year-old, my job was to brush the wonderful-smelling wood shavings off the worktable and sweep sawdust into piles on the floor.

Just as important as the suitability of the tool to the job was its relationship to the worker. The way it fit the hand, responded to leverage and force, aligned with my grandfather's thought process, reminded him of past projects or how he had inherited a particular tool from his own father, a cabinetmaker. Tools were practical, but they also told stories. They each possessed a lineage. They stirred emotions. Hanging from the rafters were hundreds of specialized jigs he had made to hold a particular part of a clock as it passed through the table saw or to route dovetail joints. As tools summoned memories, he would glance up from his work. "You know that rocking chair that sits on the porch, David?" Yes, I would nod. "Remember the legs and how they have a nice smooth bend to them?" Yes, of course. He would point to the bow in his hand. "This is what I used to form the curve."

Grandfather's tools were constructed and used with a respect for human capabilities and preferences. They fit human bodies and minds. They were a pleasure to work with and to display. They made us feel powerful, more skilled and capable than we were without them. They hung or nestled quietly, each in its place, and never made us feel stupid or overwhelmed. They were, in a word, enchanting.

WE HAVE ALREADY IMAGINED THE FUTURE OF TECHNOLOGY

I want the future of our relationship with digital technology to look less like the cold slab of glass of my nightmare and more like my grandfather's basement workshop—chock-full of beloved tools and artifacts imbued with stories. I want the computer-human interface to be an empowering and positive experience—to minimize the interruption, annoyance, and distraction of our so-called smartphones and glass-faced tablets.

Over millennia, as humans worked with textiles, wood, and metal to craft clothing, furniture, homes, and cathedrals, we developed specialized tools for specific jobs. But, in today's world, characterized by the convergence of everything into smartphones, we have become close-minded, obsessed with apps, app stores, and icons. Few innovators are daring to ask, "What other kinds of future interfaces might rival the dominance of the black slab?"

Some people, however, are imagining interfaces outside the current norm. I admire the thinking of David Merrill, my MIT colleague and founder of the inventive toy company Sifteo. He and I share a view of the needs and opportunities for human-technology interaction that are not currently being answered by the smartphone and its kin.

For one, we need to connect the billions of legacy objects that already make up our infrastructure—thermostats, doorknobs and locks, buses and bridges and electric power meters. We also need devices that can manipulate real material, such as 3-D printers that can translate electronic designs into physical objects, into food, and, eventually, into aromas. And we need tangible interfaces that make the human body smarter. Technology can enhance our five senses and optimize our physical abilities by accommodating and responding to the way we already operate in the world: with natural gestures, expressions, movements, and sounds.

What if screens atomize into a smaller, tangible, and more siftable material like sand? This is the vision of the innovative game company Sifteo. Each of these blocks is a screen that knows its orientation to the others.

These are just a few of the hundreds, thousands, possibly millions of possibilities for objects to interact with us in ways that glass slabs cannot. This book will uncover, analyze, and celebrate those objects and new forms of interaction. Technology, I believe, should help make human beings, and the world we live in, more captivating and more enchanting. You and I can help illuminate the way toward that future.

FALLING FOR ENCHANTED OBJECTS

I grew up in Madison, Wisconsin, a university town, situated on an isthmus between two large lakes. It is a town known for both its easygoing liberalism and its excellent selection of cheeses. Perhaps it was all of the sailing and boating we enjoyed, or maybe it was my father's rural upbringing, but, for whatever reason, we were obsessed with the weather. The forecast suffused the opening of almost any conversation. We regularly consulted our antique brass barometer, which hung proudly on the wall in the upstairs hall of our house throughout my boyhood and is still there today. Given to my parents as a wedding gift, the barometer is encased in brass, set in mahogany, with a white dial and two hands. You might mistake it for a clock, but if you look closer, you see that the numbers signify millibars, rather than minutes. Inscribed on the face are the words *Stormy. Rain. Change. Fair. Very Dry.* Beneath, a legend: *Falling. Deteriorating. Rising. Improving.* Every morning my father, on his way from bedroom to bathroom, would stop at the barometer, tap it, and gaze at the face as if it were a crystal ball. As he received a portent of the day ahead, he'd give a quiet "Hmmph" or "Aha" in response.

My father's weather station is inspiringly simple. It never needs an upgrade or recharging. There are no little buttons to confuse or exasperate you.

The old-fashioned barometer has come to represent for me a new and radically simpler way to think about our relationship with technology interfaces. The information the barometer had to offer could be ascertained with a quick look—it was *glanceable*. The device was polite, Zen-simple, and never intimidating. The object was dedicated to a single task of information delivery, located in one never-changing place in the house, quietly waiting to do its job. And it did so without the need for updates or upgrades or maintenance or a service plan. Our family barometer still faithfully serves my parents, nearly five decades later. The barometer came to serve as a model for me as a young interface designer, a fantastic exemplar for future interfaces. How could I make technology interactions that were this simple and convenient and useful and long-lasting?

I have always loved objects of measurement and display such as our family barometer, both real and imagined. Do you remember Frodo's sword Sting in *The Hobbit*? It's one of those fantastic objects. Not only is it perfectly made for its task—well-balanced, attractive, and sharp—it has an additional and amazing ability: it detects the presence of goblins and evil orcs. When danger approaches, Sting glows blue, anticipating its own need and use. It is a trusty weapon, an infallible warning system, a handsome object, and a fantastic companion—for a hobbit.

Sting, the barometer, and so many other steampunk-era objects— vintage car and boat dashboards, analog dials, and stereo interfaces— have material qualities that I respond to. Not only are they delightful to operate and live with, they have a knowingness about them, a possession of knowledge that they convey, an ability to amplify human abilities. Like a vintage clock, such instruments seem to carry the weight of experience.

Even as a kid, I imagined creating objects that were as handy as Sting and as mystical as the barometer. In those hours I spent in the workshop with my grandfather (avoiding the Thanksgiving or Christmas hubbub taking place upstairs), we would turn bowls on the lathe, take apart clocks, build stereo speakers and bike rollers, dream up and draw fantasy homes or airports. My curiosity carried through my childhood: at robot camp one summer, we programmed a poodle-size robot using a complicated series of codes called assembly language, and, in high school, I learned to program on my first Apple IIe, making it spin out of control with a recursive algorithm.

In college, computing opened my eyes to a new world of possibilities for what objects could already do and what they might eventually be able to do. A double major, I found that both physics and fine art had their own thrilling languages for characterizing the physical world, each with revelations and enlightenments. My graduate work at Harvard included the building of software-learning simulations like *Sim-City*. Then I came to MIT's Media Lab, a place where programmers mix with artists, musicians, and educators, and everyone experiments with technology and computation, seeking to reinvent everything from cinema to opera to medicine and education. There, I had another revelation: technology could enhance objects in ways that would come close to, or even surpass, the qualities of the magical objects from folklore and science fiction that I have loved since I was a kid. To make ordinary things as extraordinary and delightful to use and as pleasing to live with as my father's barometer and my grandfather's tools, the human-computer interaction needs to be freed from clicking and dragging. There can (and will) be real flying carpets and should be (and already are) Dick Tracy wrist communicators.

Enchanted objects: ordinary things made extraordinary.

Today's gadgets are the antithesis of Grandfather Otto's sharp chisel or Frodo's knowing sword. The smartphone is a confusing and feature-crammed techno-version of the Swiss Army knife, impressive only because it is so compact. It is awkward to use, impolite, interruptive, and doesn't offer a good interface for much of anything. The smartphone is a jealous companion, turning us into blue-faced zombies, as we incessantly stare into its screen every waking minute of the day.

It took some time for me to understand why the smartphone, while convenient and useful for some tasks, is a dead end as the human-computer interface. The reason, once I saw it, is blindingly obvious: it has little respect for humanity.

What enchants the objects of fantasy and folklore, by contrast, is their ability to fulfill human drives with emotional engagement and élan. Frodo does not value Sting simply because it has a good grip and a sharp edge; he values it for safety and protection, perhaps the most primal drive. Dick Tracy was not a guy prone to wasting time and money on expensive personal accessories such as wristwatches, but he valued his two-way wrist communicator because it granted him a degree of telepathy—with it, he could instantly connect with others and do his work better. Stopping crime. Saving lives.

The humanistic approach to computing that I propose in this book is not about fanciful, ephemeral wishes, but rather persistent, essential human ones—omniscience, telepathy, safekeeping, immortality, teleportation, and expression. To prioritize what new technologies to explore and which new devices to develop, companies and product makers must fundamentally start with human desire in its most basic forms. In doing so they can focus on creating products that can have a meaningful and positive impact in the world.

My other grandfather, my father's father, Pop Rose, died of a heart attack just after his sixty-second birthday. To my father's great regret, Pop and I never met. His death came too soon, in part because of behavioral health issues: he smoked and failed to take his heart medication regularly. He was hardly alone. As a society we are doing a better job of controlling smoking, but one of the major barriers to more effective health care, and a driver of its astronomical costs, is that people don't take the medications they are prescribed.

Today, as you would expect, there's an app for that. But, even though Pop Rose was a doctor himself and knew very well that he was at risk, would he have used a smartphone app to help him with his medication regimen? Would he have been able to find the tiny icon on the screen and use it to log his behavior? Would he have remembered all his passwords for secure Wi-Fi, iCloud, and the protected electronic medical-record system used by his doctors at the University of North Carolina?

But what if there had been a magical pill bottle—a technology-enabled object that would be as trusty as Frodo's sword, warning my grandfather that danger was lurking and urging him to take his pill? And what if that bottle had the ability to communicate with others, to let people know when he'd failed to follow his regimen?

My family's history of heart disease was one of the motivations behind my development of just such a real, "magic" pill bottle called GlowCap. It looks like a regular, childproof, amber medicine bottle, but has a special cap that glows like Sting and communicates, via the Internet, like a wrist communicator. It has enchanted users enough that people who own one take their medication over 90 percent of the time. By contrast, adherence to medications normally is in the range of 40–60 percent.

I believe that enchanted objects like GlowCap will transform the way people use, enjoy, and benefit from the next wave of the Internet—through embedding small amounts of computation, connectivity, and interaction into hundreds of everyday things that surround us, that we're accustomed to, and that have a welcome place in our homes and lives and rituals.

The idea of enchanted objects has deep roots in our childhoods, in our adulation of superheroes and fascination with fantasy and science fiction, and in the fables, myths, and fairy tales that go back centuries As a result, it seems as if we have always longed for a world of enchantment.

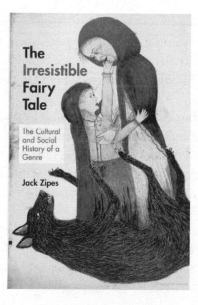

We learn about persistent human needs and fantasies from the age-old myths and fairy tales that already flow in our cultural bloodstream.

Jack Zipes is a retired professor of German at the University of Minnesota and a leading expert on the history of the Grimm Brothers fairy tales and the oral tradition that led to many of the tales by Hans Christian Andersen. When Jack heard the premise of my book—the idea that contemporary inventors should mine myth and folklore to think about the future of humanistic technology—he was hooked. We spoke at length about the origins of enchanted objects that appear again and again in stories from different cultures around the world. As you'd imagine, there are common themes:

The wishing wand or ring that fulfills any desire in an instant.
The flying carpet that swiftly transports us.
The bottomless purse that never runs out of money.
The superspyglass through which we can see thousands of miles.
Magic boots that enable us to walk miles in one stride.
The horn or whistle with which we can summon help.
The crystal ball that enables us to know the future.
The invisibility cloak or shield that hides us from danger.
The endless table that feeds hundreds with a bountiful feast.

Notice how many of these objects are transferable from one person to another. They don't provide any single person a superpower. These objects can be acquired, shared, gifted, traded, and passed down through generations—just like examples of enchanted objects I present in this book.

Futurists have speculated about the idea of enchanted objects for decades, giving the concept various names, including pervasive computing, ubiquitous computing (ubicomp), connected things, or things-that-think. The simplest, most widely used term today—usually credited to Kevin Ashton, the cofounder and former executive director of the MIT Auto-ID Center—is Internet of Things (IoT).

Arthur C. Clarke, the futurist and science fiction writer whose story "The Sentinel" inspired Stanley Kubrick's movie *2001: A Space Odyssey*, famously declared, "Any sufficiently advanced technology is indistinguishable from magic."[1] Many of today's smartest interface designers agree with him. Matt Jones, a friend and the founder of Berg, a celebrated London-based design consultancy, recently remarked, "Ubiquitous computing has been a long-held vision in academia, which has been derailed by the popularity of the smartphone." But now it seems as if we're getting closer to the Internet of Things, primarily because the price of computation and connectivity has been reduced to almost nothing.

Nearly there, but not quite. The smartphone has taken us a long way down one path, but other technology futures are vying for the attention of companies and their new-product-development dollars. These alternative visions of how technology should evolve hold great promise, but will have very different kinds of interactions with human beings and will thus deliver very different futures.

The glass slab—from the stamp-size iPod nano to the eighty-inch, ultra-HD LCD screen—now has a big lead in the technology race. I call the future defined by this sort of device Terminal World, because the interface is captured on a pixelated screen. In the early days of computing, those screens were called terminals—the "last inch" where machine met human.

For those who believe in Terminal World, such as the business leaders whose companies focus on that trajectory, the goal is to produce and distribute more and more pixels, embed screens in every surface, make devices thinner, cheaper, crammed with more features and functions, and to sell two or three to every person on the planet. Then repeat. It's not hard to envision how this scenario unfolds because it is already upon us. As of this writing, nearly 50 billion apps have been downloaded from the iTunes app store. Google's Android is rapidly catching up. And Microsoft, with its purchase of Nokia, is trying to figure out how to get in the game.

A second possible future is prosthetics—wearable technology. This trajectory locates technology on the person, to fortify and enhance us with more capabilities, to, in a sense, give us superpowers. To make humans superhuman or, indeed, "posthuman." This path of embedded wearability has some great benefits. I'm inspired, for example, when I see how prosthetics can restore physical capabilities to people who have lost them, enabling people—once considered "disabled"—to walk and run as they couldn't before, or to see or hear with range or precision they had lost or never had. However, when companies talk about a future of implants and ingestibles for everyone, I get queasy. Like plastic surgery, this future seems irreversible, fraught with unforeseen consequences, and prone to regret rather than enchantment.

An early and well-known tech prosthetic was the Sony Walkman, introduced in the 1980s, which enabled us to take music with us wherever we wanted to go and also permitted us to acoustically drop out of the world. Today's more insidious and headline-grabbing visual equivalent is Google Glass—the eyeglasslike device that projects information on a transparent screen that floats at the periphery of our visual field. The promise of this enhanced lens is that we will be able to interact with information that is displayed or projected on almost anything. While there may be benefits, risks and losses are inevitable. Walkman-style dropping out may become even more complete. You won't know when and if other people are accessing and referring to the same information that you are, or to other information altogether, or none at all. There will be no consistent, shared view of the world, even by people standing side by side. Google Glass may go even further, isolating us from each other far more completely than earbuds do today.

The third future for technology interaction is animism. In this trajectory, computers coax us into bonding with them, simulating the comforts and attraction of a living relationship. In this future, the computing intelligence is primarily located in other digital actors, not wearables or iThings. Animism stimulates the same part of the brain that gets excited by cute cats and puppy love. Animism centers on our fantasy that technology can learn us, rather than our having to learn it. Robots that could speak our language, notice our gestures, and understand what we say and wish for would unquestionably provide a pleasing human solution to the awkwardness of today's click, tap, drag-n-drop, pinch, and zoom interactions.

You probably know about the Roomba vacuum cleaner, even if it isn't yet cleaning your kitchen. The goal of the animists is to build more mobile robots of this kind, until we have surrounded ourselves with animated devices that can act as coach, butler, employee, even a friend or mate. But to expect that social robots will become a human doppelgänger, a perfect replica of personhood, is to set ourselves up for entering the zone that Japanese robotics expert Masahiro Mori calls the "uncanny valley" of creepiness and disappointment—that place where the machine's human likeness is so close to the real thing it makes us uncomfortable. Is it a human or a machine?

The most pressing question underlying these competing trajectories is this: What is the most natural and desirable—even invisible—way for human beings to interact with technology without requiring a new set of skills or constantly needing to learn new languages, gestures, icons, color codes, or button combinations? This question has fascinated me for years, driven me to start up five technology companies, and pushed me to pursue academic research and teaching at the Media Lab.

I believe these trajectories—Terminal World, prosthetics/wearables, animism, and enchanted objects—are fluid and transitional. They will all bring some degree of value and will overlap and inform each other.

I have chosen to devote my time and energy to the fourth technology trajectory: enchanted objects. I won't abandon my smartphones or lose interest in the work of my colleagues who are developing wearables and social robots. I simply believe that the most promising and pleasing future is one where technology infuses ordinary things with a bit of magic to create a more satisfying interaction and evoke an emotional response.

Think of this approach to technology as a realization of our fondest fantasies and wildest dreams. A reimagining of flying carpets, talking mirrors, protective cloaks, animated brooms, and omniscient crystal balls—as well as cherished everyday objects of our past lives, such as hallway barometers and woodworking tools—things we have always loved, dreamed about, and wanted in our lives. This book is about that reimagining and how to make it a reality.

PART I

FOUR FUTURES

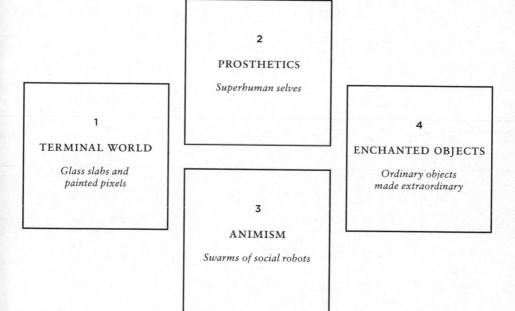

2

PROSTHETICS

Superhuman selves

1

TERMINAL WORLD

*Glass slabs and
painted pixels*

4

ENCHANTED OBJECTS

*Ordinary objects
made extraordinary*

3

ANIMISM

Swarms of social robots

CHAPTER 1

TERMINAL WORLD: THE DOMINATION OF GLASS SLABS

BEFORE DELVING FURTHER into the world of enchanted objects, let's explore the future of the other three trajectories a bit more, starting with Terminal World. Today, we find ourselves in a convergence-obsessed world, where iThings rule. How did we get here? Why has the glass slab emerged as the jealous king of things?

It is partly a matter of Newton's third law ("For every action there is an equal and opposite reaction"—in this case, when one company puts out the next-version glass slab, others respond by putting out yet another) and partly a matter of money. Making screen-based devices—from iPod nanos to smartphones to ebooks to tablets to flat-screen TVs—is a monster wave sweeping up every consumer electronics category with its massive momentum. Industry analysts, investors, entrepreneurs, app stores—the entire high-tech ecosystem—can't stop staring into the screen. Competition is intense around all the elements that make this Terminal World tick. The market for the manufacturing of pixels is staggering. Samsung, LG, Sony, and Sharp—and hundreds of suppliers who assemble components and products for these companies—are churning out millions of screens, and making billions of dollars of profit, each year.

Once you're surfing a technology wave this big, it's a huge risk to try to convince your boss or board of directors to fund anything other than another glass slab. Advocating for the next disruptive technology could mean professional suicide. As Harvard Business School legend Clayton Christensen explains in *The Innovator's Dilemma*, disruption is rarely funded by incumbents. Companies, systems, and institutions with strong vested interests in the Terminal World must invent incremental ways to gain a strategic edge on their competitors. In any mature industry, including screen-making, leaders find themselves dealing with constantly falling retail prices and are forced to compete on volume—trying to produce screens at the lowest cost and then sell them in huge quantities—or by pushing the technology forward with modest new features and capabilities and getting the new models to market faster than the others in the game. There is, right now, so much glass slab development activity that it will keep the Terminal World companies busy for years to come. The immediate next generation of products will be screens that are thinner, and remarkably larger than they are now, with more pixels per inch. Then will come organic light-emitting diode (OLED) screens with their richer blacks, and these will push past the old LED tech, which only a few years ago unseated plasma. The colors will be brighter, refresh rates higher, bezels thinner, and contrast ratios higher, so the images will look more vibrant in any kind of lighting. We will see displays in new physical forms—such as the foldable, flexible screen that you can tuck in your pocket or wrap around a building—and quantum dot (QD) displays, composed of tiny light-emitting nanoparticles that can create colors that are even more vibrant, purer, and subtler and in a wider range of tones.[1]

I spend a good deal of time with large companies (especially those associated with MIT), such as Cisco, Panasonic, LG, and Samsung. I see how challenging it is for them to shift their mind-set and pivot away from the Terminal World. When you sell pixels, it's hard to imagine anything that's not a screen. For these companies, the future of computing is not even worth debating: screens and more screens. If your billion-dollar business is selling TVs, tablet screens, and data projectors or the apps that run on them, it's hard to consider any other future, and even if you do, it's tough to figure out how your supertanker of a company could shift course to get there.

As a result, the Terminal World will continue to expand, consuming everything in its path. Not only is the market huge, with companies committed to it, but other factors will fuel its growth. The cost of pixels is constantly plummeting. Almost any surface can now be fitted out with a smart screen of some size, and the supply of information and content to pump into those displays is endless.

What's more, the amount of screen-based information that humans are capable of taking in is limitless. So, even if we're not staring directly into our smartphone or television, our peripheral vision will be saturated and distracted by dense, fast, colorful information and content that swirls at the edge of our view—as Google Glass would have it.

It's already happening. Microsoft occupies a building near my office at the Cambridge Innovation Center in Kendall Square. After being a fairly anonymous tenant in the building, Microsoft built a new, double-height entry with a big screen (perhaps thirty feet diagonally) on the interior wall, facing outward. The exterior wall is glass, so the street is dominated by the images playing constantly on that screen, totally altering the character of the neighborhood. Although the area is home to technology start-ups, big tech companies, and MIT buildings, this kind of screen domination is new.

Does the expansion of the Terminal World, even into my own neighborhood, bother me? Strangely, no. Why? Because it's hapless, obvious, and inevitable. Should we be surprised that Microsoft installed a big screen on its building to present marketing messages for all the world to see? Hardly. The screen is a blunt instrument.

Screens will continue to spread like wildfire across the landscape and into neighborhoods and places that were previously screen-free, funded largely by advertisers and sponsors seeking new ways to channel messages to affect buying habits and drive cost savings by encouraging certain types of behavior. In health care, for example, companies like United, Cigna, Humana, and Blue Shield will subsidize the pixelization of many surfaces in your home because ambient images and messages are so potent at nudging you toward a healthier lifestyle, which can lower the cost of medical care. Your living room, kitchen, bathroom, and bedroom will take on the sponsor-saturated character of today's baseball park with attention-grabbing messages displayed on every available surface. At North Station, one of Boston's main commuter-rail terminals, screens are placed almost indiscriminately, as if any space without a screen seems old-fashioned. We already see screen displays in many elevators. Expect to see more in any high-traffic public spaces such as malls and bus stops. Even the most private of public spaces such as bathroom mirrors, walls above urinals, and bathroom stalls have screens. Gas stations are following the trend, with terminals at the pump to further monetize your 180 seconds of idle eyeball time while you're filling up the tank, enticing you to step inside to buy a slushie or a high-fructose, high-margin treat. Many of these public screens also contain a camera that can sense when you're looking at it and recognize certain characteristics about you—age, gender, ethnicity, the car you're driving, the brands you're wearing—then display the products its algorithm thinks you would like the most.

Microsoft and many other companies are trying to rethink how people interact with the glass slab—with new swiping gestures, for example—but it's still all about screens. To see just how screen-centric the culture of Microsoft is, and how the company imagines the future, watch some of its online videos that present the company's vision of the future. (You can find links at enchantedobjects.com.) You will see that screens running the Microsoft Surface user interface will be available in every size and shape and customized for every context, from schools to airports to museums and bedrooms. They show palm-size versions, tabloid formats (for traditionalists who still like the idea of a "newspaper"), screens as big as a desk or, better yet, large enough to fill your wall.

This vision of the future can hardly be called a vision at all because it offers nothing new. It just extends the familiar and obvious line forward: same thing, different sizes, different places. For businesses, the Terminal World future is a snap to imagine, the way forward is relatively clear, next-quarter results are falling-off-a-log simple to forecast, and by sticking with it you avoid any disruption to your product-development plans. Your career is safe.

But here's my gripe with black-slab incrementalism. Screens fall short because they don't improve our relationship with computing. The interfaces don't take advantage of the computational resources, which double yearly. The devices are passive, without personality. The machine sits on idle, waiting for your orders. The Terminal World asserts a cold, blue aesthetic into our world, rather than responding to our own. Even the Apple products, celebrated for their hipness, are cold and masculine compared to the materiality of wood, stone, cork, fabric, and the surfaces we choose for our homes and bodies. Few of us long for garments constructed of anodized aluminum with a super-smooth finish.

The Terminal World does not care about enchantment. The smartphone does not have a predecessor in our folklore and fairy tales. There is no magic device I know of whose possessor stares zombielike into it, playing a meaningless game, or texting about nothing. It does not fulfill a deep fundamental human desire in an enchanting way.

CHAPTER 2

PROSTHETICS:
THE NEW BIONIC YOU

TODAY, TWO HUNDRED thousand cyborgs are walking the planet—cybernetic organisms composed of organic and nonorganic parts. You may not notice them as they pass by you because they look like rather ordinary human beings. But these creatures have surgically implanted computers, first developed in the mid-1960s, that connect directly to their brain—more specifically, the auditory nerve in the inner ear. Cochlear implants and the benefit they bring are miraculous. A person born without hearing who receives a cochlear implant can recognize, without lip-reading, 90 percent of all words spoken and 100 percent if they do lip-read. This is the benign, positive, even magical prospect of the second future: prosthetics and wearable technology.

This future has its technological antecedents in the fantastic worlds of comic books and imagination: superhumans and mutants, bionic men and women, the unbelievably powerful and swift and capable. Unlike Terminal World, the prosthetics future for technology *does* take into account our humanity. Prosthetics amplify our bodies, the power of all of our senses, and the dexterity of our hands. It's appealing to develop technology that keeps us more or less who we are, only more so. We already have memory, and the technology gives us much, much more of it—fantastic storage and retrieval capacity. A Google-like brain. It's important for technologists to understand this desire to have superhuman powers and extraordinary abilities, to be able to fly like Aladdin or Peter Pan, to leap tall buildings at a single bound like Superman, or to be able to see through walls and around corners like Peepers, the Marvel mutant with telescopic and X-ray vision.

The critical characteristic of technology-as-prosthetic is that it internalizes computational power. It becomes a part of us, so much so that it *is* us. It's not out there, external, captured on a screen, forcing us to do something to activate it. The vision of prosthetics is like the cyborgian man or bionic woman. *The Six Million Dollar Man* was a popular television show in the 1970s based on the novel *Cyborg* by Martin Caidin. The hero, former astronaut Steve Austin, has been severely injured in the crash of his flying vehicle. Six million dollars later, Austin has a replacement arm, two new legs, and an eye upgraded with a high-precision zoom lens. He can run as fast as a car and lift enormous weight, and his eye is just as sharp as a magic telescope.

The spin-off show, *Bionic Woman*, features tennis pro Jaime Sommers. After she is seriously injured in a skydiving accident, Sommers bounces back with better legs, an ability to jump to great heights, and superhearing. Even with these bionic parts, Steve and Jaime look like normal human beings. They are improved versions of themselves, rather than Frankensteins or quasi-robots.

This is the fantasy of the bionic person. We remain fundamentally human, but technology-hacked. We look and behave normally, but we are better able to see, hear, remember, communicate, and defend ourselves than the standard all-human model. No wonder bionic people in fantasy and popular culture typically manifest in such roles as secret agent, explorer, or soldier. They can parachute into any environment, peer through darkness, anticipate every hazard, ford raging rivers, scale daunting cliffs, effortlessly kill (and cook) the next meal, construct a shelter and thrive—all single-handedly. I find this entertaining but also limiting. Wouldn't it be great to see a bionic person who does something other than spy or fight? I want to see a bionic musician, inventor, architect, or city planner! What might those kinds of cyborgs achieve with their enhanced abilities?

In addition to having enhanced or special powers, many superheroes rely on prosthetics. Superman's archrival, the mad genius Lex Luthor, wears an exoskeleton that makes him stronger and less vulnerable to injury. In the Batman and Robin stories, Clayface sports an exoskeleton that enables him to melt people. I find this vision of prosthetics particularly unappealing because it doesn't tap into basic positive human desires such as omniscience and creation. These technologies are tools of violence, revenge, and madness—and they're also clichéd.

What amplified abilities and superpowers do real people crave today? Noise-cancellation technology to drown out the din of the world so we can concentrate better. The ability to detect free Wi-Fi zones and their bandwidth. A mechanism to turn off the annoying TV at an airport or jam the cell phone signal of a yappy fellow traveler. In a cacophonous world we often don't want to see and hear more than we already do; rather, we want better filters so we see and hear *less* or just exactly what we want. We don't want augmented reality, we want diminished reality. That is the modern version of the ancient wish to have superability to subdue the things that threaten us.

THE HEADS-UP DISPLAY: AUGMENTED VISION

The consumer prosthetic technology of the 1980s and '90s was auditory, but in this decade we'll see the visual equivalent of the Sony Walkman and the iPod—the personal heads-up display, or HUD. Today's HUDs are embedded in glasses or goggles, onto which information is superimposed and appears to float in the air before your eyes, such as with Google Glass, or in a large glass field, such as a car windshield.

A better design of these systems is urgently needed. The HUD in the car, for example, lies directly in the driver's line of sight. The benefit is that you don't have to look down at an information display and take your eyes off the road, but the risk is that you get distracted by the info-clutter on the windshield. You hit cognitive overload, lose focus on the road, and fail to react to a deer crossing the highway or a car veering into your lane. The challenge for engineers and designers is to create a HUD that is bright enough to be seen on a sunny day, but not so bright that it overwhelms when driving at night. Carmakers are thinking of HUD as a way to make their products different, to stand out from competitors'. Not only will it display the standard dashboard information—speed, gas consumption, audio selection, and the like— but also information from the Internet. Expect the windshield HUD, which is still expensive, to become standard not only in autos but on many other glass surfaces such as conference-room walls, doors, even bus windows. In military applications, where the information is dense and subsecond performance is critical—and cost is no barrier—a HUD is the standard in aircraft cockpits and weaponry. (With a self-driving car, of course, you won't need a HUD, just a head-down pillow for you to sleep on. More on that later.)

So what's the problem with projecting information onto everyone's glasses to provide a personal lens on the world? At least three factors— beyond the obvious one, cost (outside the military)—have prevented wearable, personal HUDs from taking off. First, the devices have been too large and uncomfortable to wear continuously. Second, they are still so ugly that any reasonably self-respecting person wouldn't choose to wear them. Third, the information they present isn't useful enough to outweigh their potential for distraction. Is it really important to con- tinuously monitor your tire pressure or washer-fluid level as you drive along? No. An occasional glance does the trick and is probably safer. If you're looking through a display continuously, the information design needs to be much more subtle than those envisioned in any movie fan- tasy of robocops, superspies, or iron men that I have seen.

I'm interested, however, in early-adopter categories in which work- ers—such as military pilots—*must* wear glasses and helmets. Commer- cial aircraft and helicopter flight crews are also natural candidates for information-delivery prosthetics of some kind. Construction work- ers don't care if their gear is fashionable, and, most important, the information they could access could be incredibly valuable to their work and safety. Expect HUD technology to move beyond military and industrial uses and become a factor in consumer activities, particularly performance sports such as motocross, hang gliding, skiing, and snowboarding. In 2012, Oakley, the eyewear maker, introduced the Airwave ski goggle, fitted with an accelerometer, a gyroscope, GPS, and Bluetooth. The information is displayed on the goggle lens in real time and can also be accessed later, so you can analyze your speed and other data. The goggles also have the advantage of keeping you from getting lost when a snowstorm closes in or you can't find the trail. Professional athletes, such as football players and NASCAR drivers, are also candidates for HUD gear. One day the quarterback, instead of consulting the scrawled plays on his armband, may see the next play flash up on the visor of his helmet.

To gain widespread popularity outside of such specialty applications, the HUD will have to become more attractive and wearable and significantly enhance one or more of our abilities. Human memory is the best candidate. It is now so fallible and fleeting, I can see the value in a prosthetic HUD service that would deliver real-time information in a wide range of situations. Imagine a social setting where you don't know all the people or a business meeting where you have not been fully briefed on the topics under discussion. At the party, wearing your fashionable HUD display, you will instruct the device to display the people's names and key biographical info above their heads. In the business meeting, you will call up information about previous meetings and agenda items. The HUD display will call up useful websites, tap into social networks, and dig into massive info sources such as the National Digital Public Library (NDPL), the portal to all of the holdings of every archive, library, museum, and university in the country.

This kind of memory-enhancing prosthetic will provide the equivalent of a supercharged brain. No matter what the topic, you will have access to the figures, the references, and the opinions. You will fact-check your friends and colleagues (although this might have an effect on your popularity). You will also engage in real-time messaging, including videoconferencing with friends or colleagues who will participate, coach, consult, or lurk. This will be useful when you need to consult an expert—when, for example, you're in the middle of a tricky surgery, or trying to master kiteboarding, or exploring a new city, or need shopping advice. The market for virtual-vicarious entertainment (already with us in the form of reality TV) will be expanded.

One of the earliest prosthetics, the monocle, only caught on when it became a fashion statement.

The adoption of wearable devices will be accelerated as technology blends with fashion. In the latter part of the nineteenth century and into the early twentieth, a distinctive prosthetic device—the monocle— soared in popularity, blown by the winds of fashion. Its function was to enhance vision, but it also made a trendy statement. It became associated with wealthy, upper-class men and, when worn with a morning coat and top hat, completed the image of the properly outfitted 1890s capitalist. Similarly, Google Glass will gradually become more fashion-focused than it is functional in the next decade as its capabilities become more useful and better integrated into our lives.

Fashion has helped many a new technology move from technical device to popular accessory. President John F. Kennedy helped sunglasses make the jump from function to fashion. The pocket watch evolved to the wristwatch. For each of these mature wearable technologies, think about how many variations and designs exist today. There are thousands of watch brands and models and sunglass designs, and the market for design-differentiated headphones is booming. Fashion differentiates and variations proliferate.

My point is that prosthetic and wearable technologies have a different adoption pattern and criteria from external technologies such as screen-based devices. With prosthetics, fashion dominates technology and functionality. It is one thing to slip a cover on your phone, quite another to replace a wristwatch your father gave you or forgo your comfy and stylish glasses for a clunky helmet.

A DISAPPOINTING INTRODUCTION
TO GOOGLE GLASS

Will Google Glass be the breakthrough HUD, the one that becomes a fashion bellwether and must-have technology? As I was working on this book, I met with a special project group called Google X in Mountain View, California, Google's home base. The Google X team is full of crazy-smart MIT people, many of them friends—including Astro Teller, who focuses on wearables as well as artificial intelligence, and Richard DeVaul, also an expert on wearables and formerly an engineer with Apple. This team is responsible for the Google Glass project, the self-driving car, and other moon-shot ideas about which I'm sworn to secrecy.

Before I tell the story of my date with Glass, let me say that, as a researcher and innovator, I love trying breakout ideas and I forgive the failings of first versions and early iterations. I can see past the lack of polish, stuttering showmanship, and even the occasional mid-demo reboot. It's the integrity of the idea that counts, even if the demo only lasts a few minutes before going up in smoke. It's the glimpse of the inevitable future that is exciting. So I went into my trial run with Google Glass with an open mind. I'm a bleeding-edge early adopter and wanted to walk away wearing a fashion-forward prosthetic that would forever change the way I interact with technology. Throw away the phone, the iPad, the flat screen, and the laptop. Just give me Glass!

Even before I got to the campus, I was fantasizing about how Google Glass might transform my life. I pictured myself driving along Highway 101 to San Francisco. The driving directions would appear on the road ahead of me. As cars approached, the ones whose drivers had worrisome driving records would have polka dots superimposed on them as a warning. As I came into the city, an augmented-reality layer would identify buildings that had once stood there before being destroyed by earthquake or fire, and also a vision of how planners and architects envisioned the city as it might look in the future. As I passed people walking their dogs on the street, I would see their names (humans) and breeds (dogs) hovering above them. When I stepped out of the hotel to eat, Google Glass would offer advice on dining places in the area and the types of food they offered, recommending items on their menus. When I went for a run the next morning, a famous runner would be running alongside me virtually, setting the pace and encouraging me.

In other words, I was pumped. I entered the Google X building wanting to believe that the time for personal heads-up displays was here, that Google had made a bold investment in the future, and that this team had developed the next big thing. The team's space was a pleasing surprise. It was uncluttered, cubicle-free, and had none of the appurtenances and playthings so common to start-up "theater"—slides, foosball table, and beanbag chairs—but looked and felt more like a big, open warehouse. So far, so good.

The big moment arrives. I'm handed a Google Glass prototype. Before I slip it on, I inspect it. It is so light I guess it is made of titanium. (It is.) The distinguishing feature is the small screen, about one centimeter square, built into the frame and positioned about a half inch in front of my right eye. I slip the device on, and my dreams are instantly shattered.

This is not an augmented-reality experience, it is not a moment of enchantment—it is just another version of glass-slab Terminal World. A little screen floats in my vision, as if it were hanging a couple of feet above me and just to the right of the Google guy sitting across the conference table from me. The little screen doesn't actually block anything important, but it's uncomfortable to have to keep looking up and to the right while I'm trying to have a conversation.

To wake up Google Glass—to bring its functions alive—I am required to say, "Okay, Glass." (Inspector Gadget comes to mind:

Go, go, Google Glass!) A cute little ding sounds in my ear, and on the screen, a menu (oh, no, please, not a menu) appears. The options include Video, Google, Messages, Directions. I say, "Video," and the screen displays a shot of the room I am in—the scene I am already looking at, with no enhancement. After ten seconds of recording, GG shuts off. Nothing much of interest has happened in that time. Continuous recording, I am told, quickly fries the battery, which is housed in the temple piece of the glasses. I understand the trade-offs the designers are grappling with. They're working to find the optimal configuration of function and fashion, computational power and battery life, size and weight. They are favoring lithe and small, which is the right choice given that all wearable displays have so far been too large and too ugly to wear continuously the way you wear regular eyeglasses. Even the Google Glass I tried that day was too large and silly-looking to wear for long. My Google friends weren't worried about these limitations, however. "Give it three to five years," they said. "With a thirty percent reduction in size and weight each year, we'll get it where it needs to be."

Maybe. But even if Google Glass sheds 70 percent of its bulk and can run for days on a single charge, two critical user-interface issues remain to be solved: input and augmented reality. To control Glass today, the interface is a combination of voice and touch. You speak and stroke the temple of the glasses in one of four directions: up, down, forward, backward. The speech control is awkward and doesn't work well, if at all, in noisy environments. It doesn't augment your ability to speak, but rather diminishes it.

The interface issues can be worked out, but the real promise of Google Glass is to be a genuine augmented reality (AR) experience. That is a much harder challenge to meet. To accomplish it, Glass will need to recognize and understand precisely where your gaze is directed and superimpose relevant information on the object, landscape, or person you're focusing on—and make it visually stick. When you move your head, the information needs to stay connected to the object or landscape or person, rather than move across the screen in the same position. Think about it: none of the rest of the world travels with your head, except for the smudge on your glasses.

Once Google Glass is able to superimpose information with a "world reference," the real fun begins—but another problem crops up: the "filter bubble." If I am living in a bubble that is filtered differently from the one you see, it is as if you and I are living in two separate worlds. Wearing the iPod has the same effect in the auditory domain. We can't talk about the music because you can't hear what I hear. Relativistic visual views will be even more isolating.

This lack of a common view is hugely problematic but also conceptually interesting. What happens when we filter the world differently? Let's say I want to see the eBay price of every car on the street, obscure the brand logos on the clothes people are wearing, and peer into the history of every built structure I encounter. You, on the other hand, want to know the price of the shoes people are wearing, learn what coupons are available at stores you pass, and see movie times and previews on movie posters plastered on the sides of buses and cabs.

Does that matter? Aren't we already doing that kind of filtering in our minds? Don't we all see the world from very different perspectives? Yes, but when Glass or its equivalent is working at full strength, it will be as if you and I are tuned into different radio channels, and the channels are all-encompassing. The experience may divide and isolate us.

Is there a better way? One idea would be to include the capability to tune into each other's channel. For live events such as sports and concerts, it would mean you could see every view from every seat. And if I could spend an hour "walking in your shoes" by "looking through your glasses," what an opportunity for empathy! It would be eye-opening and bring people into greater engagement on issues such as poverty and politics. It would enhance my understanding of others, rather than reduce it through a hyperfiltered, hyperpersonalized view of the world. Of course, even as we enhance our understanding of other human beings through technology, technologies will also be trying to understand human beings better. Which brings me to the third technology trajectory—enter the social robots!

CHAPTER 3

ANIMISM: LIVING WITH SOCIAL ROBOTS

EVEN THOUGH PEOPLE often ask if he is crazy, Dmitry Itskov has set what he believes is a realistic deadline—the year 2045—to complete a project the *New York Times* described as nothing less than the "mass production of lifelike, low-cost avatars that can be uploaded with the contents of a human brain, complete with all the particulars of consciousness and personality."[1]

Currently the 2045 Initiative is mostly in the mind of Itskov, a young Russian oligarch, but it has also begun taking tangible form—the form of his own head. A robotics company is creating a reproduction of Itskov's head and face, which will be animated by thirty-six tiny motors to produce humanlike expressions. Uploading the full contents of Itskov's mind will be the more difficult challenge. Undeterred, Itskov sees the production of fully functioning and totally sentient human avatars as a way to alleviate many of the world's problems, including world hunger and anxiety. "We won't need shelter anymore," Itskov told *Forbes* in a 2013 interview. "We won't need to consume resources that we consume now. This body won't depend on food. Health care would be focused on the repairing of the new artificial body and not the biological system."[2] And because the entire contents of your mind would be transferred into an infinitely reparable robotic creature, it would fulfill the wish to live forever.[3]

The fantasy and lure of animism—the replication of life—persists in the efforts of Itskov and many other futurists. Might this trajectory eventually be the primary way we interact with technology and services? Even as screens proliferate and prosthetics for sight, sound, and touch are more available, will we continue to want to relate with technology possessing humanlike features? I think so. That's why the idea of the social robot is so compelling, but is, like the smartphone, a dead end as a human-machine interface.

IN OUR OWN IMAGE

If I say the word *robot*, what do *you* picture?

A humanoid with quirks, right? A little awkward, like C-3PO, the bashful robot from *Star Wars*, or his predecessor Maria, from Fritz Lang's 1927 film *Metropolis*. You imagine a device of more or less human scale (not building scale or cockroach scale). The robot of your imagination understands human speech (and your particular language and accent) as well as the meaning of gestures, and it can speak and gesture in response to you. The robot is also mobile, probably bipedal, just like you.

The word *robot* is attributed to Karel Capek, the Czech author and playwright, from his 1920 play *Rossum's Universal Robots* or *R.U.R.* According to the *Oxford English Dictionary*, the word derives from the Czech *robota*, meaning "forced labor" or "drudgery."[4] In Old Russian, the word *rab* means "slave."

The robot is one of our oldest and most durable visions of how we will interact with technology. The idea of artificial humans—and other types of faux creatures—goes back to the ancient Greeks, particularly to the early engineer Heron of Alexandria, who claimed to have developed a steam-powered mechanical bird, known as an aeliopile, sometime in the first century of the first millennium. (Heron is also credited with developing the first coin-operated vending machine.) Aristotle showed interest in the idea of robots when he mused that it would be handy if every tool could "accomplish its own work" by "obeying or anticipating the will of others." If that could happen, then "chief workmen would not want servants, nor masters slaves."[5]

Leonardo da Vinci likely constructed the first robot, sometime around 1495, in the form of a knight in armor. According to Mark Elling Rosheim, author of *Leonardo's Lost Robots*, this knight "sat up; opened its arms and closed them, perhaps in a grabbing motion; moved its head via a flexible neck; and opened its visor, perhaps to reveal a frightening physiognomy."[6] Leonardo's robot was remarkable in its workings. It consisted of "an elegantly simple differential pulley/cable drive system" embedded in its breast, in which "the cables exit the robot from the back or base" and "power for the robot to move its arms and stand or sit came from manual operation or perhaps a waterwheel."[7]

In the modern era of robotics, we have struggled with the concomitant ethical and moral dilemmas that surround the merger of human and machine. The science fiction writer Isaac Asimov tried to come to grips with these issues by defining, in his 1942 story "Runaround," the Three Laws of Robotics:

1. A robot may not injure a human being or, through inaction, allow a human being to come to harm.
2. A robot must obey the orders given to it by human beings, except where such orders would conflict with the First Law.
3. A robot must protect its own existence as long as such protection does not conflict with the First or Second Laws.[8]

Developers of robotic devices have not always faithfully adhered to Asimov's laws. Drones, which are the most widely used and technologically thrilling type of robots today, violate all of Asimov's dictums. They are designed to kill or injure human beings (Law 1), obey orders to harm human beings (Law 2), and protect their own existence (Law 3) so that they can keep breaking the first two laws.

In movies, however, robots are often favored characters, and they tend to be Asimov Law–compliant. C-3PO, of *Star Wars* fame, is a benign helper, a Swiss Army knife of assistance. Other humanoid robots play on the question of empathy and identity and our desire to remain as special beings—in possession of souls—and thus distinct from machines. For example, we fear becoming mechanical versions of Pod People, the soulless duplicates created by visiting aliens in the movie *Invasion of the Body Snatchers*.

The border between humans and machines can be titillating, creepy, and horrifying. Ridley Scott's sci-fi cult classic movie *Blade Runner*, based on a story by Philip K. Dick, features a class of humanoids called replicants, which are human clones who have been created by the evil Tyrell Corporation to work for people who populate space colonies. The replicants are in many ways physically superior to humans, but they do not have authentic emotional responses. They are designed to have a fixed, four-year life span and, because they are so humanlike and violent, are relegated to spend their short existences on off-world colonies, prohibited from visiting our beloved planet. When four of the most advanced model—the Nexus-6—mutiny and escape their space colony to find their way to earth, a blade runner, played by Harrison Ford, is brought out of retirement to terminate them. (The movie is set in 2019. We don't have much time. . . .)

One of the replicants, a gymnastic punk-android named Pris ("a basic pleasure model"), is taken in by J. F. Sebastian, a genetics designer employed by Tyrell Corporation. We discover that Sebastian has created hundreds of replicants for his personal use, mostly to keep him company, and also to use as sex partners. In the movie *Her*, Spike Jonze portrays a vision of a deeply personal relationship between a human and an operating system. The operating system is never directly embodied, but only worn as a hearing aid or headset. These movies offer a vivid vision of a world in which humanoids have distinct personalities and play a role that, though not quite human, are deeply social. Sebastian has a little tribe of flawed and unfinished friends. Their quirks make them distinct as machines and also adorable, precisely because they have imperfections and need human guardianship just as babies do.

NEOTENY: THE POWER OF CUTENESS

The attractive attributes of babyness figure prominently in animism. Product designers are always thinking about how people will interact with objects and services, especially those that will be used daily. Will people bond with them? Form an emotional attachment with them? One of the tricks designers use in creating robots, toys, and hundreds of other products is to play on the human response to neoteny—cuteness—which largely results from the inclusion of "juvenile characteristics," such as big, unblinking eyes, dilated pupils, an oversize head for the body.[9]

That our response to neoteny is almost always positive has been proved by researchers in academia and business, particularly marketers. Byron Reeves and Clifford Nass report on this effect in their book *The Media Equation: How People Treat Computers, Television, and New Media Like Real People and Places*. The authors find that we have an uncontrollable habit of mapping human characteristics onto inanimate objects if they show any signs of life. In *The Man Who Lied to His Laptop*, Nass posits a thought-provoking theory on their Stanford web page. "Our brains can't fundamentally distinguish between interacting with people and interacting with devices. We will 'protect' a computer's feelings, feel flattered by a brownnosing piece of software, and even do favors for technology that has been 'nice' to us. All without even realizing it."[10]

Human beings attach to objects and imbue them with lifelike personalities. It doesn't take much for us to start thinking of and interacting with an object as if it were human. Give a device a blinking LED, a hiplike curve, a smile-shaped grill, and we start to impute personality to it. It seems trustworthy, mercurial, arrogant, friendly. When a machine can move around, our expectations for intelligence and personality are further sparked. "Throw in a face," as Sherry Turkle, the MIT sociologist, says, "and we are goners."

One small example supports this theory: hundreds of blogs report on people's attachment to their Roomba vacuum-cleaner robots. Perhaps it's the smilelike curve on its top, combined with its dumb but deliberate determination—bumping into one thing after another but never daunted—and the way it spins and emits a continuous mechanical purr that make Roomba too cute and charming to resist.

Automotive designers take this tendency into account, especially in the design of the "face" of a car. Look at the evolution of the BMW 3 Series over the years. What kind of "person" is the 1968 model? With its wide-open eyes, big open mouth, petite[11] grille-nose, bare feet, and high forehead? And what kind of personality has emerged by 2006? Somebody more guarded, grown-up, less friendly?

Given our susceptibility to the slightest hints of personality, it's easy to extrapolate that our future interactions with machines and services will be ever more infused with personality and that we will increasingly form relationships with these devices, even if they are not in the robot tradition. We already make judgments about many other nonhuman objects—tools, toys, household appliances, cars, even buildings—and interpret that they are smiling, frowning, showing anger or delight. I expect that we will have hundreds of machine-based relationships in our lives, each with a specialized role: tutor, shrink, coach, shopper, jokester.

I asked Sherry Turkle specifically about the MIT robot Kismet, designed to help in the research of these issues around social robots. Kismet has emotive eyes with huge pupils, long lashes, actuated lids, and a gaze meant to mirror a human's—more precisely, a child's. Kismet attends to what you attend to. The effect, Sherry admits, is overwhelming. Sitting across from Kismet, you quickly attach to it. You are soon reading the emotion and intention in Kismet's face, inferring goals, sensing that the machine genuinely cares about you.

THE RULE OF RECIPROCITY

Robots like Kismet raise an essential question: What does social interaction with personality-possessing technology do to us? The social psychology concept (and commonsense understanding) of reciprocity holds that the way we behave toward others influences the way others behave in return. This principle underlies human conduct. One good deed begets another. But does the rule of reciprocity apply to how we interact with robots?

Again, Stanford's Clifford Nass has useful research to contribute. In 1996, he investigated how people would react toward robots that seemed to be reacting to them in particular ways. In his experiments, people were much more willing to perform a tedious exercise for helpful computers than for nonhelpful ones. Nass's findings support the argument that we develop humanlike relationships with nonhuman objects.

Researchers have devised other ways to get at the nature of the relationships between humans and robots. Christoph Bartneck, for example, tested the robot version of Milgram's obedience study, the famous electric-shock experiment where the subject ratchets up the voltage until it's clear the person receiving the shock is suffering—but, at the urging of the researcher, keeps right on going. (The shock is not real.) Bartneck's experiment involved a cat robot. The subjects in the experiment were asked to play a computer game while a cat robot sat by their side. Some of the cats were helpful to the players; some were not. Once the people finished their games, they were instructed to power down the robots. The twist: the cat robots did not want to be shut down and began to plead for their lives. Understandably, the participants were disturbed. They felt as if they were killing the cats. The people who were assisted by helpful cat robots were especially reluctant to press the off button.

FALLING INTO THE UNCANNY VALLEY

Why is all this significant? Considering our tendency to respond to neoteny, and taking into account that we are so easily beguiled by any machine that displays a hint of a personality, we reach an essential problem with social robots: the theory of the uncanny valley.

Although we've always wanted a replicant, a doppelgänger, a robot made in our image, the closer we get to reaching that goal, the less pleasing it becomes. If we could create a perfect human doppelgänger, one that had all of our communication skills, logic, memory, and emotional sensitivity, that would solve the problem of human-computer interaction nicely. We wouldn't have to call it Human-Computer Interface. The machine would be sufficiently human that, for all intents and purposes, we would be designing a Human-Human Interface.

THE INEVITABLE ISSUE WITH ANDROIDS

The uncanny valley theory predicts that we will reject almost-human robots

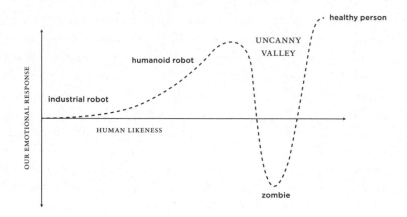

But building a human facsimile has proven elusive. In 1970, robot researcher Masahiro Mori coined the term *uncanny valley*. His insight is that as a machine gets closer and closer to humanness, the likeness becomes so good that any imperfection becomes unacceptable, even creepy. We have no problem with an industrial robot, or a C-3PO, because they are clearly not flesh and blood. They don't threaten or confuse our own sense of self. Frankenstein, on the other hand, approaches a human facsimile. He is mostly constructed of human parts. He walks, talks, and seems sentient. Thus, his nonhuman components—the bolts in the neck, stitches in the forehead, and lurching walk—make him disturbing. Mori concluded that it was foolhardy to create robots that attempt to exactly mimic humans.

Though powerful, Mori's warning has not stopped the animists. In 2000, Honda unveiled a robot it calls Asimo, "the world's most advanced humanoid robot," and it teeters on the edge of the cliff that overlooks the uncanny valley. Asimo's movement is extraordinarily human. It hops on one foot and runs at a brisk 9 kmh with limber knees and a spring in its step. It picks up a bottle, twists off the cap, and pours water into a glass. It walks among humans without bumping into them, shakes hands, and gestures naturally with its arms and fingers. It's almost as if a small human being were tucked inside a plastic suit, except for those large ball joints connecting Asimo's torso to its legs. What keeps Asimo on the machine side of the valley is largely the face. There is none: just a blank, dark panel.

Asimo is an excellent example of "category confusion." We don't like to be baffled by the nature of the creatures around us. That's why we have a hard time dealing with gender ambiguity (remember Pat, the androgynous character on *Saturday Night Live*?). It's why most women and men take care to distinguish themselves from one another through clothing, hairstyles, makeup, stance, and gesture. Androgyny threatens our stereotypes and triggers category confusion and anxiety.

The salient question is, would you want Asimo living in your home? It (Asimo is gender neutral) takes up as much space as a person (even though it doesn't bump into you), but is not capable of deep human interaction. It can perform tasks that seem amazing for a robot but elemental for a person and would generally not be helpful around the house. Yet the experience of interacting with such a humanlike "living" robot is rich and compelling enough that it challenges our sense of human specialness. It makes us feel anxious. Asimo's almost-humanness evokes feelings of mistrust, competition, and embarrassment. As human as it looks, we don't know how it will behave. All our years of human observation are rendered useless. Is it foolish to shake its hand? If we get too close, might it kick us in the shin with its foot-like slab of hard plastic?

An advanced humanoid robot, an android, is a human being in everything but the most important elements: heart, mind, soul. The uncanny-valley theory explains why we are more likely to bond with a Muppet than we are with a bad version of a human (who can relate to a CPR dummy?). The poor version of human form is unforgivable. A robotic dog or a cute dinosaur such as the PLEO is not only forgivable, it can be enchanting. The future of the robot will be in delightful toys like Furby, or dogs like the Sony AIBO, or task-specific robots like Roomba, the floor sweeper, and Scooba, the floor-washing robot.

HOW MANY AGENTS DO YOU WANT?

An important consideration for interaction designers and businesses that hope to build engaging and useful robot-based services, each with a personality, is the structure, number, and hierarchy of the relationships between human being and machine. Would you prefer to have many "agents" and distribute the work among them? Or would you prefer a single agent that can perform many jobs and services or delegate them on your behalf?

The popular PBS drama *Downton Abbey* illustrates the distributed-agent model. The show chronicles the lives and times of the wealthy Crawley family in the early twentieth century. The Crawleys employ a large number of servants, who work inside the huge Jacobean-style country house and also look after the grounds, farms, and stables. Each has a specific role: butler, footman, cook, housekeeper, housemaid, scullery maid, chauffeur, groom, gardener. Would we want to re-create such a world, but with animated devices taking the place of Anna, James, Carson, and Mrs. Hughes? Or, to use the modern equivalent, do we want many Siris in our life?

As the cost of robotic services and replicants goes down, this becomes a design and lifestyle question similar to "How many pairs of shoes will I have or teapots will I collect?" Of course we would love to get more help with all kinds of tasks: to clean house, teach us new skills, coordinate transportation, handle legal affairs, plan vacations, manage money, provide health-care assistance, look after the kids. We want a masseuse, a personal shopper, even a proxy therapist. Most of us already use, or want to use, some or all of such services, but they are too expensive or burdensome or inconvenient or time-consuming. As costs go down and we get better at creating pleasing, task-specific robots, however, more people will be eager to off-load work onto devices.

What degree of animism will we want in these service providers? Should they have personalities like replicants, be humanoid like Asimo, or relatively mechanical like Roomba? I believe we will want our service providers to be "personalitied"—to have just the right amount of human character and hyperspecialization so we don't feel threatened or overshadowed and can remain in the comfortable role of generalist.

I certainly want my *human* service providers to be personalitied. At the gym, I look for trainers whose demeanor and interaction style appeal to me. A trainer has to hold me accountable, be willing to listen to my kvetching about being sore and popping ibuprofen, and empathize with my goals. I am not motivated to go to the gym if I am only going to interact with the machines, no matter how much data they might be able to display about my workout.

A service provider with a personality is able to activate motivational systems that are innate, depend on social relatedness, and are, as a result, powerful reinforcements. The trainer with a personality can activate unconscious childhood transferences. *She will approve of me. She will love me if I do well.* We are driven by the wish to please our parents and by the romantic partner fantasy. These kinds of reinforcement systems are always at play in our relationships and are at work in the "companionate" systems that fire up our mirror neurons—the brain cells that help us empathize with others, even if we have little or no direct connection to them. (This is why we cry at movies.) These innate systems are powerful, and animate-style human-computer interaction will inevitably leverage them.

THE PERSISTENT PROBLEM WITH ANDROIDS

Given the advancements in robot technology and the tremendous promise they hold for a new kind of human-machine interaction, why haven't social robots emerged as the primary way we interact with technology?

My friend Cory Kidd from the Media Lab launched a company called Intuitive Automata, to bring to market a social robot whose sole purpose is to help people with dieting. Autom is personable and knowledgeable, and unlike Roomba and other robots, the product is designed and identified as female. Autom takes her place in your kitchen, perched on the counter next to the fridge. She talks to you about food in general and your eating in particular. She knows the calorie count in almost every food, offers motivational comments, and makes diet suggestions tailored to your characteristics—every day—all for a price of less than a single therapy session.

Autom sits in your kitchen and watches what you eat. Will you eat healthier because of your relationship?

Autom has promise, but interacting with her does not take full advantage of our desire to relate to robots as we do humans. Autom has a human look, but in the place where her belly should be is— oh, no—a flat glass screen. The Teletubbies Terminal World intervenes once again. Autom communicates only through text displayed on her screen. Worse, you have to touch buttons to interact. The effect is of an iPad holder with a neotenic head featuring big eyes and a perma-smile. Is it a robot with a belly screen or a smartphone with a cute face?

Autom may be effective for weight management because she's knowledgeable and stationed right next to the fridge, making her hard to ignore when you reach for forbidden foodstuffs. Ultimately, however, Autom is disappointing because she doesn't recognize or talk to you or have the ability—like Kismet—to look at you with her eyes and react to what you're stuffing in your mouth. Maybe the next iteration will solve these problems and add other capabilities, perhaps with connections to other sensors that can help you track and manage physical activity, sleep quality, stress levels, and your daily schedule. That would give Autom a much greater understanding of who you are and what you need to do to trim down and health up.

Autom provides a good example of how incredibly difficult it is to create robotic systems that are believable and useful. They're brittle. They often fail. Robots with speech recognition often can't understand accents or get confused by ambient noise. Face recognition often fails in low light. Learning algorithms often need to be untaught and then retaught.

Many of these failures are hastened by the user. When people encounter a new technology, they like to test its boundaries, just like a four-year-old child, trying to probe its capabilities and limitations of themselves and the world. When evaluating a new car design, test drivers will veer sharply from the guardrail on one side of the track to the opposite guardrail to test how quick and responsive the steering is. Designers of all kinds need to consider this "guardrail experience"— what are the outer limits of the product's performance? The designers at Apple anticipated that people would want to test Siri in this way. When you say, "Siri, I love you," she replies, "I value you." "Siri, how much do you weigh?" you ask, and she quips, "I thought this conversation was about you." Siri accommodates these oral gibes in a clever and charming way, so, even though you know she has no personality or backstory, you accept her level of humanness—just enough to be relatable, but not too much to be creepy.

As we get better and better at robot design, it will be hard to distinguish the humans from the humanoids. Sherry Turkle believes that machines should not look human or appear charming precisely so they will not activate human emotions, and thus the distinction between humans and machines will be preserved. I disagree. I understand that category confusion provokes anxiety, that people become disturbed when they're not quite sure if they're talking to a machine or a person. But I also believe that natural human-computer interaction has more benefits than risks.

When I was in high school, the standard breakup line was "Let's just be friends." The snide retort was "If I wanted another friend, I would have gotten a dog." Given Facebook's and LinkedIn's abilities to connect me with hundreds of loose ties (aka "friends"), I don't feel the need for a new set of artificial friends in my life, especially in the form of awkward, sometimes dim-witted, and often brittle robots.

I'd rather have better and more enchanted tools.

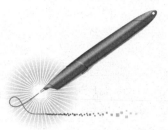

CHAPTER 4

ENCHANTING
EVERYDAY OBJECTS

WE'VE TALKED ABOUT locating the future of human interface in black slabs or terminals, on our bodies as wearables or prosthetics, and outsourced to a tribe of social robot slaves. The fourth future—the one that captivates and occupies me—is enchanted objects.

Enchanted objects start as ordinary things—a pen, a wallet, a shoe, a lightbulb, a table. The ordinary thing is then augmented and enhanced through the use of emerging technologies—sensors, actuators, wireless connection, and embedded processing—so that it becomes extraordinary. The enchanted object then gains some remarkable power or ability that makes it more useful, more delightful, more informative, more sensate, more connected, more engaging, than its ordinary self. As the ordinary thing becomes extraordinary, it evokes an emotional response from you and enhances your life.

Recall the magical objects from myths and fables that Professor Jack Zipes described as parts of our common cultural bloodstream. The replenishing purse that constantly pours forth money, the horn that can summon help across vast distances, boots that enable you to run without tiring, the carpet that can fly. Such objects continue to animate our movies, games, comics, and popular fictions. Enchanted objects are the real-world manifestation of these fabled desires.

These wallet prototypes provide tangible feedback. One becomes harder to open as you approach your monthly budget. The other puffs up so you can feel a recent electronic deposit.

An everyday object that gains magical powers is both captivating and comfortable. First, it becomes easy to relate to, like the two-way wrist communicator. We know about watches and radios. The form is familiar. We have an instinct about how to use them, they have a place in the space we already inhabit. Their purpose coincides with our daily goals. An ordinary-looking wallet holds cash and credit cards, yes, but an enchanted wallet also has a wireless link to your online bank account so it becomes harder to open when you're going over your budget, overusing credit, or making purchases on impulse.

An enchanted shoe is not only functional and fashionable, it records your steps, keeps track of your pace, analyzes the stability of your gait, and motivates you to exercise.

The light from an enchanted lightbulb subtly shifts color as your home energy usage increases or decreases, prodding you toward resource conservation.

Take the ordinary pen that I'm using to write this sentence. What would make it enchanted? A bottomless ink supply? Real-time spell-check? Infinite memory to capture every word? Knowledge of the day, the café, and the person with whom I'm speaking when I take notes? The Livescribe pen, created by the company Anoto, comes close. It's a writing instrument augmented with a camera, a microprocessor, and a wireless connection. It becomes a fantastic tool that captures your notes as you write them, stores them digitally, and transfers them to your computer or iThing. The device works with a special paper printed with a fine dot pattern, which enables it to recognize and record the time and position of each pen stroke.

The Livescribe takes the enchantment further by adding audio recording capability. When you listen to a lecture or conduct an interview with the pen in hand, the audio is recorded, and every time you jot down a note, the timing for each letter is stored. If you want to check what was actually said at the point in the conversation when you were only able to scribble a ?, just tap the pen on the ? and the recording plays from that moment forward. This capability makes the audio far more useful later because you can instantly find the salient snippet you need without having to search through minutes or hours of conversation. It also impacts how you take notes because you can use a more symbolic language or your own form of shorthand. While researching this book, for example, I used stars, smiley faces, and hearts to categorize certain types of points during interviews. I could replay a good bit just by tapping the symbol in my notes and the pen would play it.

If I knew that *you* had the Anoto Livescribe while reading this book, I'd scatter little glyphs on the page for you to tap to hear the voices of my interviewees as you read the quotes. (The book could also have a sound track, with great songs to hear while we talk about certain aspects of enchantment.) This technology could allow us to recast our expectations around the ebooks experience and think beyond Kindles and tablets. The ebooks of the future could feature enchanted objects such as pens or wands that talk or a sound track for each page. While you read the novel of the future, your home data-projector might show an image of a forest, just as you read the words *they entered into a deep woods*.

The Livescribe pen achieves a nice fusion between analog and digital while preserving all the familiar characteristics that make a pen such a pleasing tool. It looks like a pen, works like a pen, but is much more than just a pen. Although you love it for its extra capabilities, its essential "pen-ness" isn't compromised.

DESIGNING THE FUTURE WITH TAPE

It can take some time for people to understand and fully appreciate the many ways that objects as familiar and simple as a pen can be enchanted—not to mention the multitude of responses enchantment can provoke. At MIT I teach a graduate course called Tangible Media. Each semester we work together to develop prototypes of enchanted objects or services with tangible interfaces. By *tangible* I mean that the interaction between human being and object does not require a screen. Instead, interfaces rely on gesture, tactility, wearables, audio, light and pattern, and haptics—the use of touch.

Over the semester, as students sketch different ideas, they are always struck by the potential of enchanted objects—they begin to see just how useful and convenient computing will become as it sweeps through the world of things. They grasp how these non-screen-based interfaces will change the fabric of the ways in which we live, work, and play. A jacket that gives you a little hug when someone likes your Facebook post; a paintbrush that samples any color or pattern you tap the brush on and turns it into digital paint; tables that listen, furniture that melts into the floor or into the wall when it's not needed; lights that understand your activity and adjust their intensity and focus appropriately; watches that help students meet other people like them and prompt face-to-face conversation; E Ink Post-it notes that dynamically update to show place-based messages; a key fob that displays the traffic situation on your commute ahead. Those are just a few of the great ideas from the past few semesters. The possibilities and prototypes keep pouring out.

To help students open up their thinking, I lead an improvisational activity early in the semester. I give each student a piece of ordinary masking tape. Their task is to stick it to any object that might provide an enchanted service and to imagine any kind of interface, sensor, or display that might be embedded in the material of the object. It's a generative, expansive exercise, and it sparks an outpouring of interesting questions about the object, the imagined service, and the potential interface, such as:

- Would the display be public or private? Could anybody see the information the object provides, or only me, the user?
- How dynamic is the data? Would the data change and update constantly in real time or at some longer interval?
- What is the trade-off between distraction and value? How intrusive is the object, and is that intrusion acceptable, given the value the object delivers to me?
- Is the display encoded or camouflaged or easily learnable? How much effort does this object demand of me in the interaction?
- Does the interface mimic (or biomimic) the natural world in some way, such as wilting? Does it pick up on something I already do, such as a gesture or an action?
- What's the agency and personality of the service? How does the object represent me? Is it cute? How do I feel about it as a presence in my life?
- Is the information mirrored or can it jump to other touch points such as mobile, Web, or large public displays? Can I get the information it provides in other ways and places or is it confined to that one object?
- What's the learning curve? How quickly would I be able to use it effectively?

THE MACRO TRENDS FOR ENCHANTMENT

Enchanted objects may be just one of four interface trajectories, but its arrival is inevitable. My research predecessors, such as Mark Weiser at PARC, as well as my current colleagues at MIT, including Hiroshi Ishii and Neil Gershenfeld (who authored *When Things Start to Think*), have watched the macro technology trends—miniaturization, embeddedness, and, importantly, cost—and come to similar conclusions.

The cost of computation, sensors, and displays moves downward in a nonlinear way, and nonlinearity is hard for humans to comprehend. We see the effect of nonlinear trends in compounding interest, flu pandemics, and viral videos—but we don't yet have an intuitive sense of them in computing. Speed of computation doubles every eighteen months, but more profoundly—especially for the Internet of Things and enchanted objects—the cost for the same amount of computation is cut in half. As a result, the cost of processing has become 128 times cheaper in the last decade, so cheap that it is almost an insignificant consideration in the development of new connected products and services. We can embed silicon and sensors in any object—shoes, pill bottles, lightbulbs, wallets, and furniture—virtually without noticing the incremental costs.

Is there an object that can't or won't be enchanted? You could suppose that the enchantment of real-world objects today, unlike in our fantasies and folklore, has its limits. But if you look at the trends in miniaturization, wireless networks, and costs, you will quickly see otherwise.

People who do not believe in ubiquitous computing and a worldwide Internet of Things make several arguments against the enchanted future. The first is that the cost will still be too high for general adoption and that the complexity of execution will be prohibitive. They concede that the makers of big-ticket items such as cars and refrigerators may be able to embed any additional cost of enchantment into the selling price and no one will notice or object. But, they argue, the same will not be true of less expensive objects like teapots and toasters, wallets and lightbulbs. In many product categories, consumers are keenly aware of cost and search for the best price they can find.

I doubt these doubts. Technologists have already shown that the price thresholds of today will be history tomorrow. So long as people delight in enchantment—in its utility, simplicity, and wonder—the cost of technology will vanish as a barrier to enchanting things.

INGESTABLE-SIZE SENSORS

Designed by Sonny Vu, the Shine tracks walking, biking, and swimming in a quarter-size piece of titanium.

Critics similarly argue that enchanted objects have size limits. You can only go so small, they say. Surely some objects aren't big enough to accommodate sensors, wireless communication, circuit boards, and batteries? But here again, technologists are constantly upending expectations and overcoming size barriers. Jawbone, Nike, Polar, and Fitbit offer bracelets with embedded sensors that track physical activity. The company Misfit offers a titanium tracking brooch called the Shine, and Narrative of Sweden developed a clip-on brooch with embedded camera and GPS that takes a photo every thirty seconds, creating a movie of your life. At Vitality, we prototyped a smart credit card, two millimeters thick, that can track physical activity and provide points and bonus dollars to people who exceed a walking goal they have set for themselves. Proteus offers ingestible circuits embedded in pills to track and assure medication adherence. Miniaturization will not be a barrier to enchantment.

This credit card contains a battery, microprocessor, and accelerometer to provide financial incentives for getting enough exercise.

THE TECHNOLOGY HALF-LIFE PROBLEM, ADDRESSED THROUGH MODULARITY AND THE CLOUD

Doubters further argue that the value of enchantment depends on the rate at which products age and lose their appeal or utility. Does it make sense to enchant objects we know will go out of fashion quickly (such as clothing)? Wouldn't that be a waste of design energy and resource? What about objects that we buy and expect to use for many years (such as furniture or train cars or jewelry)—does it make sense to embed a rapidly changing technology in things that hardly ever change? Wouldn't that be a route to disenchantment? Who wants a ten-year-old piece of furniture encumbered with a technology that became obsolete three years after purchase?

The auto industry has been dealing with this issue for years. I saw this clearly when I spent a three-day weekend at the Epcot Center in Orlando, at the invitation of a large car company. I was one of a group of experts assembled from an intriguing number of fields, including luxury travel, restaurant design, NASCAR racing, and materials science. One expert researched new types of glass that dynamically change opacity or absorb sound. Another was a PhD in aerospace from MIT (a real rocket scientist). The work of this Lead User Group was facilitated by a team from the MIT Sloan School of Management, led by Eric von Hippel, one of the world's leading authorities on innovation. Von Hippel believes that diverse groups of early adopters who come from outside the organization can quickly generate breakout ideas. Our mission was to come up with new ideas for automotive design.

Over the three days, as we came up with hundreds of great ideas, we kept bumping up against the reality that each part of the car has a different half-life. The electronics system is out-of-date before you buy it. Some content, such as maps and music, needs continuous updating. Floor carpets may wear out after one salty, slushy winter. Cloth seats, especially when cracker-happy, juice-spilling kids are involved, need periodic refreshment. Mechanical elements such as shock absorbers, boots, and seals can last for five to ten years. The metal skin of the car gets scratched, fades, and rusts after a few decades in a harsh climate such as we have in New England, but the chassis could be sound for a hundred years or more. So we kept searching for ways to manage these differing life spans in the architecture of the car's components.

A solution to the life-span problem—and for many kinds of enchanted objects—is a modular architecture. That is, architecture in which individual components can be updated, upgraded, and swapped in as needed. In the automobile, for example, a constant connection to the Internet means the dynamic content of music and maps is always up-to-date. Furthermore, a connection to the cloud allows "upsourcing" of computationally complex tasks such as speech recognition and route planning that take into account weather and traffic congestion. The cloud will play an essential role in upgrading the features and functionality of the car through a technology called over-the-air updating (OTA).

Not long ago, I saw compelling evidence that OTA will be an important player in creating enchantment that lasts. One day just before Christmas I was rambling through the mall and noticed a group of gawkers at a Tesla car dealership. I strolled over to have a closer look at what was captivating them so. A new Tesla S, the all-electric "family car," was on display. The exterior styling is reminiscent of a Maserati's, except the Tesla accommodates six people (you can opt for a third row of two small, rear-facing seats) while only two can squeeze into the Maserati. I slid into the driver's seat and looked into a dashboard fashioned entirely of glass. A seventeen-inch touch screen controls media, climate, navigation, and other functions. (I said earlier that the four technology futures will converge and overlap—Tesla is creating an enchanted object that also offers the benefits of Terminal World.)

A salesperson explained the car had an amazing function we could not see: it could be improved and updated while parked in its owner's garage. We wanted to hear more. The salesman offered an example. One Tesla owner had asked for a "crawl feature," which would enable the car to creep along in slow-moving commuter traffic. No problem. Tesla did an over-the-air update that created a new accelerator function—delivering just enough power to keep the car moving—and placed a new check box on the preferences page on that giant seventeen-inch dashboard screen. The owner was happy, and now the crawl feature is available to all owners and future owners who want it. Check the box and the car will crawl forward at a speed of a few miles per hour without your foot on the accelerator.

The OTA capability has a tremendous impact on Tesla's business model because it affects how people think about their car purchase. If I can download new features and behaviors through software updates from the cloud, I should be able to keep the hardware version longer than I might otherwise have.

Two things will happen as a result. First, OTA updates will be marketed and priced more aggressively. *The Tesla update package for 2017 is available now for $4,999. Just say "authorize" to bill the credit card on file.* Second, there will be more pressure on designers to vary and differentiate the physical look of the car from year to year. If the behavior or function of the car (the software) is more readily downloadable and available, then the impetus for upgrading, and the way to communicate prestige for having the newest iteration, will be through form, not function.

For cars, and other devices, being connected to the cloud has other, and bountiful, secondary effects. It enables greater serviceability. The car or washing machine will know what's wrong with it and connect with the service facility to fix it, if possible. Cloud connection makes updates easier and even automatic, so there will be no need to make a call to the service provider to upgrade your objects. And connection increases "findability" (a LoJack system for everything), as well as sustainability. Why discard the hardware if you can independently update and improve an object's behavior?

ENCHANTED OBJECTS ARE AVATARS FOR SERVICES

The cloud—along with miniaturization, ubiquitous connectivity, and falling costs—will be essential to the trajectory of enchanted technology. In addition to OTA capability, here's another powerful way to think about enchanted objects in relationship to the cloud, courtesy of my friend and colleague Mike Kuniavsky, the PARC researcher.

Mike says that connectedness allows an object to become an avatar—a physical stand-in, a tangible manifestation—for services, which are otherwise intangible. Think about the American Express card. It's an object with a magnetic strip and embedded radio frequency identification (RFID), but it is actually an avatar for your secure spending power and a symbol of prestige. It facilitates access to services such as concierge support, travel insurance, access to airport lounges, data encryption, yearly reporting services, and connections to other touch points such as Web portals and useful apps. All objects should be so well endowed.

To develop objects as avatars for an ever-expanding basket of services is a dream for hardware makers. They are always looking for ways to produce revenue after the initial sale of the product has been made. A subscription. A tether. A warranty. A membership. Enchanted objects, when they act as avatars, provide companies a way to stay connected to their customers and offer them a steady stream of new capabilities—and create new streams of revenue, as well.

The business and service implications of enchanted objects will impact industries like no other macro force in this century. Remember when AOL carpet bombed the mailboxes of America with a free CD full of content, games, and a thirty-day trial for AOL service? The 5 percent conversion rate AOL achieved—acceptance of the offer for a $9.95-per-month service with an average total customer lifetime value of $350—justified what seemed like a crazy effort and huge expense.

A persistent connection like the one AOL achieved (for a time anyway) gives a company the often exclusive opportunity to service, up-sell, cross-sell, upgrade-sell. Think razors and blades. Software companies that have no cost of goods are blazing the way with models such as the "freemium"—they give away the basic software but charge for new or additional features. It's a freemium because the product is free at first but encourages you to pay a premium later. This is an example of playing the long game—the company invests money to expand its user base, which it then monetizes as time goes along.

Sooner or later, and probably sooner, Amazon will transform Kindle into an avatar for services rather than continue to market it as a Terminal World object—or just another black glass slab. The Kindle is an ideal Trojan horse for bringing Amazon's vast catalog of books, movies, and products into every home, purse, and backpack in America. It's a media pipeline in which Amazon can control the content *exclusively*. Amazon should give the Kindle away at every bookstore in the world, Velcro them to textbooks, stuff them into Lucky Charms cereal boxes.

Connectedness through the cloud and the ability to transform objects into avatars will affect every kind of company, including apparel makers like Nike and Ralph Lauren, retailers like Staples and Walmart, household and personal product makers like P&G, Colgate, and Estée Lauder, as well as transportation providers, and of course pharma and insurance companies. In the next several years, companies throughout the Fortune 500, operating in every category of goods and services, will be looking for—and will find—ways to enchant objects in such a way that they will act as avatars for a wide range of services.

Not all of them will be successful. People will resist the attempts to enchant some objects and embrace others. Why? What will make the difference?

The enchanted objects that will succeed will be the ones that carry on the traditions and promises of the objects of our age-old fantasies, the ones that connect with and satisfy our fundamental human desires. They will be cars that transport us as safely and as delightfully as flying carpets, writing instruments that remember, rings that connect us, tools with as much utility, familiarity, and character as my family's barometer.

PART II

SIX HUMAN DRIVES

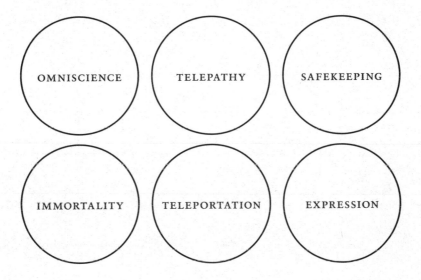

OMNISCIENCE

TELEPATHY

SAFEKEEPING

IMMORTALITY

TELEPORTATION

EXPRESSION

THE DIALECTIC INTERPLAY:
FICTION AND INVENTION

In *The Wizard of Oz*, the house where young Dorothy lives with her aunt Em and uncle Henry on the Kansas prairie is engulfed by a violent cyclone, whirled around three times, and lifted like a feather to the very top of the funnel. By some anomalous combination of weather and magic, the little house—with Dorothy and her dog, Toto, soaring along inside—is transported to the Land of Oz. There the cyclone loses steam and the house comes hurtling down, crash-landing on a little old woman, who turns out to be the Wicked Witch of the East. All that is left of her is a pair of silver shoes, conveniently poking out from under the corner of the newly transplanted house. The residents of Oz, the Munchkins, are delighted with the demise of the Wicked Witch and assume Dorothy must be a good sorceress. They insist she take the Witch's silver shoes because, as one Munchkin says, "The Witch of the East was proud of those silver shoes, and there is some charm connected with them; but what it is we never knew."[1]

Only later does Dorothy learn, from the Good Witch, that the shoes do have extraordinary powers. "And one of the most curious things about them is that they can carry you to any place in the world in three steps, and each step will be made in the wink of an eye. All you have to do is to knock the heels together three times and command the shoes to carry you wherever you wish to go."[2] That's exactly what Dorothy does to return herself (and Toto) home to Kansas. In the movie version of the story, the silver shoes were upgraded to ruby-encrusted slippers—so they could pop and glitter in Technicolor.

Moviegoers have been so enamored of the idea of a pair of shoes that can transport you wherever you want to go (while also looking terrific) that the costume slippers from the 1939 film have become one of the most valuable pieces of movie memorabilia. In 2012, the pair that Judy Garland wore when she clicked her heels (five pairs were made) sold for an undisclosed sum to a cabal of Hollywood foot fetishists led by Leonardo DiCaprio, who donated them to a museum planned by the Academy of Motion Picture Arts and Sciences. Before the auction, the shoes' value was estimated at $2 to $3 million.

The 1950s fantasy of playing Scrabble rather than driving is coming back, thanks to Google.

That is enchantment. An ordinary object. Something we already know how to use. Imbued with a fantastic power that fulfills one of our most fundamental human desires—to travel effortlessly to a place where we feel safe and natural: home. Dorothy's ruby slippers, which may have been inspired by mythic characters such as wing-footed Hermes or Mercury, are antecedents to a number of real-life enchanted objects now appearing—with extraordinary capabilities. Nike, for example, has made a significant investment in a range of wearables—including Nike+ shoes, which have an accelerometer embedded in their soles to measure your pace.

I call these dialectics—pairs or series of objects, each one inspired by the ones that came before (sometimes long before), connected to the same fundamental human desire. I use the term *dialectic* because the creation process is a kind of dialogue—encompassing and embracing diverse and often divergent points of view—across cultures, over time, and among and between creators of fiction and fantasy and makers of real-world objects. Hermes's sandals begat Dorothy's slippers begat Nike's Nike+, all of them promising to transport us to a place of our dreams, but employing different approaches in form and style and delivering different takes on what the dream might be.

We will see this dialectic again and again in the enchanted objects I discuss, and I hope that more of our current inventions explicitly follow this pattern, taking their inspiration from the imagination that infuses our mythologies, fairy tales, and pop culture. Countless new technologies and product prototypes never get out of the lab (or garage) or, if they do reach the market, don't take off. Many more do come to market and become part of our lives but fail to enchant us. They serve a purpose and even become ubiquitous (parking meters) but are cumbersome, confusing, and inelegant. They are tech things that we tolerate or use out of necessity, but they fail to spark our imaginations and engender our love.

The mirror from Snow White has inspired another narcissistic mirror for comparing outfits at Neiman Marcus and sharing with your Facebook friends for fashion advice.

The experiences that do enchant us reach into our hearts and souls. They come from the exotic place of "once upon a time." They help us realize fundamental human desires. The fantastic technologies we have invented over the centuries, the ones of ancient tales and science fiction, enable us to do things that human beings earnestly want to do but cannot do without a little (or a lot) of help from technology. They make it possible to fly, communicate without words, be invisible, live forever, withstand powerful forces, protect ourselves from any harm, see farther and travel faster than the greatest athletes. They are tools that make us incredible, supercapable versions of ourselves. These are the visions and stories of our most beloved authors of fiction and fantasy—Tolkien and C. S. Lewis and J. K. Rowling and the Grimms—and the realities of fantastic characters such as Cinderella, Dick Tracy, James Bond, Superman, and Wonder Woman.

The designers creating enchanted objects must, therefore, think of themselves as something more than manipulators of materials and masters of form. They must think beyond pixels, connectivity, miniaturization, and the cloud. Our training may be as engineers and scientists, but we must also see ourselves as wizards and artists, enchanters and storytellers, psychologists and behaviorists.

THE ESSENTIAL PSYCHOLOGICAL DRIVES

Which are the most fundamental, universal, and timeless human drives? Which most influence our behavior and actions? Which desires do technologies tend to serve and satisfy?

Let's take a brief tour of some of the origins of our thinking about things. In ancient Greece, Aristotle (384–322 BC) believed much of human behavior is driven by desires, such as the craving for pleasure or self-preservation. "It is manifest . . . ," he writes in *De Anima*, "that what is called desire is the sort of faculty in the soul which initiates movement."[3] But desire is not the sole driver, he says. It is balanced with reason, and together they influence our decisions and actions.

Heed these ancient words! An enchanted object must strike a balance between practicality and pleasure, form and function. It must not be purely practical so as to be boring or so pleasurable that it only satisfies the hedonic desires.

Thomas Hobbes (1588–1679), the English philosopher and pessimist, believed that aversions and fears drive us.[4] Charles Darwin (1809–1882) focused on the instinct to survive as the most fundamental motivator. And Sigmund Freud (1856–1939), reclining on his Austrian psychoanalytic couch, conjured the id, which was all about primal instincts, including the sex drive and the fear of death.

In the twentieth century, American psychologist Abraham Maslow (1908–1970) proposed a framework for structuring human desires that is still widely in use. In his 1943 paper "A Theory of Human Motivation," he proposes we all have five essential needs—physiological, safety, love and belonging, esteem, and self-actualization. Humans, Maslow argued, must satisfy the needs in an ascending hierarchy starting with survival and self-preservation, then climbing to the pinnacle of self-actualization.[5]

Maslow's hierarchy is useful for technologists and designers to keep in mind, so they can be clear about what need their invention addresses. Are you satisfying the need for safety or helping people self-actualize?

I'M AN INTJ, WHICH IS LOGICAL

In college I took a psychological test that was in vogue—the Myers-Briggs Type Indicator—and I learned from it that I am what is known as an INTJ. The Myers-Briggs classifies people according to four psychological functions: sensation, intuition, feeling, and thinking. These functions set up four dichotomies that determine one's type: "extraversion or introversion (E-I), sensing or intuition (S-N), thinking or feeling (T-F), and judging or perceiving (J-P)."[6] According to the MBTI, everyone has tendencies along these scales, and these four tendencies comprise your type. For example, someone who is extroverted relies more on sensation for information gathering, makes decisions based on feelings, and has a lifestyle predicated more on judging than perception would be typed an ESFJ. I'm an INTJ, which in practice means that I recharge when I'm alone and am more likely to make decisions based on logic than emotion.

The Myers-Briggs framework is frequently used by employers to improve workplace communication and dynamics, but it's difficult to quantify its effect.[7] I know people who find it helpful in understanding themselves and those around them and go so far as to introduce themselves, "I'm an ESFJ, what are you?" Perhaps the fusion of your Myers-Briggs classification, combined with insights gathered from the analysis of Big Data (along with your Klout score), will result in some better way to more deeply understand human motivations. We'll see.

I have not found such typing and classifying frameworks—or even the slicing and tabbing of huge data sets—to be particularly helpful in dreaming up, designing, developing, and taking to market new products and services. In my work as a product inventor and entrepreneur, as a teacher and as a consumer, I have thought a lot about what makes products resonate with certain audiences and what fundamental human behaviors are at work behind the scenes. I've synthesized my observations, as well as the thoughts and research of many others, to identify a set of human desires that I believe are fundamental and universal, and that deserve the focus of product designers and entrepreneurs and companies.

- *Omniscience.* This is the desire to have great knowledge. We have a voracious appetite to know as much as possible and to know about things that go beyond facts and information. We would love to be able to predict what will happen in the future.
- *Telepathy.* We have a powerful desire to connect to the thoughts and feelings of others, and to be able to communicate with ease, richness, and transparency. We want to know others and to feel known by them.
- *Safekeeping.* We fervently wish to be protected from harm. To feel comfortable, safe, and at ease.

- *Immortality.* We want to be healthy, strong, fully capable. We dream of living long lives, vital to the last moment.
- *Teleportation.* We crave movement, to be transported easily and swiftly and joyfully from one place to another, and to live unconstrained by physical limits or boundaries.
- *Expression.* We all wish to be generative, to fully express ourselves in many forms and media—acting, music making, art, writing, cooking, dancing, documenting our lives.

In this part of the book, I will delve into each of these in depth.

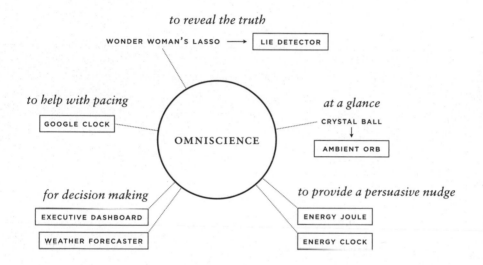

to reveal the truth

WONDER WOMAN'S LASSO ⟶ LIE DETECTOR

to help with pacing

GOOGLE CLOCK

at a glance

CRYSTAL BALL
↓
AMBIENT ORB

OMNISCIENCE

for decision making

EXECUTIVE DASHBOARD

WEATHER FORECASTER

to provide a persuasive nudge

ENERGY JOULE

ENERGY CLOCK

OMNISCIENCE: TO KNOW ALL

WILLIAM MOULTON MARSTON (aka Charles Moulton) belongs to a rare breed. He is one of the few people in history who created an imaginary technology and then invented the real thing—a one-man enchanted-object dialectician. Marston is known both for his role in developing a key component of the first lie detector test, the Marston Deception Test,[1] and as the person behind the camp comic book heroine Wonder Woman.

Born in 1893, Marston was a Harvard-trained psychologist, a champion of the causes of women (whom he believed, based on his research, had better characters and were better workers than men), and led an unusual private life: he lived in a polyamorous relationship with two women. Perhaps it is not surprising then that Marston predicted in 1937 that women would form an "Amazonian matriarchy," wage a "battle for supremacy" with men, and, in a millennium, "take over the rule of the country."[2] This hypothesis is literally and metaphorically manifested in *Wonder Woman*, which debuted in 1941 in All Star Comics #8 (and has been published by DC Comics ever since), in which Amazonesque women live on "Paradise Island" with eternal life and happiness—without men.[3] Marston believed that women were less violent and materialistic than men and could thus "lasso" them with sexuality.[4] Wonder Woman's chief weapon is her Lasso of Truth, or Magic Lasso, with which she can coerce people to tell the truth.[5]

Wonder Woman's Lasso of Truth, a gilded band forged by the god Hephaestus, used by the heroine as either a noose or a whip, is a fantastically enchanted object: when anyone is in its circle, that person must provide any information he or she is hiding. It is as if Marston wished he had his own lasso of truth, so eager was he to come to a more profound understanding of human nature.

Marston was more in love with his fictional character and her truth-telling device than he was with the lie detector test he helped develop. (As a doctoral candidate in Harvard University's psychology department, Marston studied systolic blood pressure and created a way to measure it in intervals. The modern polygraph, developed by John Augustus Larson in 1921, is a modification of Marston's test that provides continuous blood-pressure readings.) Marston wrote in the *American Scholar* that "the picture-story fantasy cuts loose the hampering debris of art and artifice and touches the tender spots of universal human desires and aspirations, hidden customarily beneath long accumulated protective coverings of indirection and disguise." Comics, he said, speak directly to "the innermost ears of the wishful self."[6]

Marston saw that our appetite for information is insatiable, especially for information that is distant or secret. The same wish has excited and spurred technologists and futurists for centuries. Sundials and clocks enabled us to see and mark time; astrolabes and compasses gave us position and direction; thermometers and barometers reveal the temperature and pressure of the air, through which we can forecast the near future.

Today, we are deluged with information, data we cannot directly know, about the state of the world. We scatter such information everywhere, wear it on our wrists, display it on our bedside tables. We spew it out in streams of Tweets, news feeds, blog posts, and other media. But much of that information is bland and of little value. It masquerades as news, but in its form and delivery, it lacks enchantment. It's just there—in overwhelming quantities.

OMNISCIENCE AS WIZARDRY

The earliest technologies for accessing information combined the technology of the day with elements of magic and wizardry. Two thousand years ago, in the Chinese Han dynasty, the first compasslike device featured a spoon-shaped pointer made of material with magnetic properties that rested on a "diviner's board" inscribed with elements from the *I Ching*—the ancient Chinese text also known as the *Book of Changes*—and symbols for the stars and planets. The device was used for prediction and prognostication—"to determine the best location and time for such things as burials."[7] The perspicacity of such a device is dubious, but it is easy to imagine how its use filled observers with a sense of wonder.

Over centuries of storytelling, we have devised all manner of devices to enable us to find answers to our deepest and most urgent questions: *What will happen in the future? What is the condition of the person I love right now? Which path should I follow?* What's wonderful about fantastical objects in fiction, like all enchanted objects, is that they are typically based on the everyday technologies we have come to understand and trust, but with magical features and added capabilities. They extend a known object with a tantalizing extra dimension. Some of the most intriguing, like the golden compass and the magical clock in the *Harry Potter* books, seem ancient but have been imagined in modern times. We may love our screens and smartphones, but we are still attracted by the mystical aspects of the kinds of devices that might be used by druids and wizards.

THE TRUTH-TELLING COMPASS

Where is omniscience found? How have we imagined it? What devices, surfaces, or mediums have we thought capable of making us omniscient? I suggest looking at two of the wildly popular series of fantasy literature of recent times: Philip Pullman's fantasy trilogy, His Dark Materials, and J. K. Rowling's Harry Potter series.

Pullman imagines a device called the alethiometer, also known as the golden compass, which combines the analog hands of a clock with features of a compass, blending the real with the imagined, the practical and the mystical. The word *alethiometer* comes from the Greek word *aleth*, meaning "truth," so you might call the alethiometer a truth-o-meter. It will tell you the truth about anything you ask, deliver the answer to any question, no matter how complex and no matter how supported or unsupported by available data it might be. *Should I take this job? Should I marry this person?* In *The Golden Compass*, Lyra, Pullman's protagonist, often uses the alethiometer to reveal what is hidden—objects lost, the location of people around the world.

The alethiometer is complex, its weight manifests its quality of craftsmanship, and it is beautiful to behold. The dial is strewn with tiny icons, little images—including a horse, an owl, an anchor, a moon, a mandolin, a bee, a globe, and a lightning bolt. As Pullman writes, the pictures are "painted with extraordinary precision, as if on ivory with the finest and slenderest sable brush."[8] The alethiometer, according to Pullman's story, is supposed to have been devised in the seventeenth century, a time when "symbols and emblems were everywhere. Buildings and pictures were designed to be read like books. Everything stood for something else; if you had the right dictionary, you could read Nature itself."[9]

The user interface of the alethiometer, however, is a nightmare—more complicated than a Ouija board or tarot cards. Not just anyone can use the device; it requires an alethiometrist, who may train for years to become proficient in the craft. To pose a question, the alethiometrist moves the three clocklike hands so they point at certain icons in a particular sequence. Once the hands are positioned, a longer, fourth hand magically darts to several different icons, sometimes flicking back and forth among them several times. The alethiometrist intently watches the movements, arranging them in his mind, and then formulates based on what he has seen. The interface is hardly natural and the information is not glanceable, but the device is enchanted because it conjures up knowledge we could not otherwise divine. It is omniscient. One caution, however: do not alethiometrate and drive.

THE ORIGIN OF THE AMBIENT INFORMATION ORB

I was admiring the features of familiar objects, such as house clocks and compasses, and the simplicity of interacting with them, when I came up with the idea for the Ambient Orb.

Inspired by magical crystal balls and the philosopher's stone, I chose the shape of a pebble for the glowing Ambient Orb.

I love old clocks. And, as I've confessed, I am hooked on old weather stations and barometers. They are always there, politely waiting for you to notice them. Unlike phones and computers, clocks respect your attention. They aren't interruptive. They have a calm presence. They don't require you to do anything to gather information from them—you don't have to find an icon, launch an app, or type in a URL. They don't push their information at you, unless they have chimes or cuckoos. They are there, always available, in every room of the house, with the exact information you expect from them. And—unlike the black or silver consumer electronics aesthetic—clocks are differentiated by fashion and designed to complement our living environments in countless pleasing ways.

As I was looking for a form that could display information in a home-friendly and aesthetically pleasing way, I was also inspired by a prototype information lamp, built by a student in the Tangible Media Group at the MIT Media Lab, led by Hiroshi Ishii. Andrew Dahley and Craig Wisneski borrowed a ripple tank from the physics lab, positioned it above a spotlight, and hooked up its perturber (the moving armature that causes the ripples) to visit a website. The more people visited the website, the more ripples appeared on the tank's surface, creating concentric circles thrown onto the ceiling. People could instantly perceive the ripple shadows in their peripheral vision. As opposed to text or numbers, it was cognitively lightweight.

Another inspiration for a tangible, motion-based information display also came from Hiroshi's group. An origami paper pinwheel, a popular children's toy in Japan, is spun by a tiny motor whose speed is determined by its connection to the Web. As the number of visitors to the associated website increases, the pinwheel spins faster. It is as if a breeze of digital bits is invisibly moving the paper pinwheel. It's delightful and magical—allowing you to see the unseen. It succeeds as an ambient display. It's faster to "read" information from movement than text. Still, motion in the periphery can be distracting.

My eureka moment came when I realized that I could express information not as movement in water or pinwheel blades, but as subtly changing colors. Why not program a trend in information to a shift of hue along a spectrum? A person would quickly come to know the color range and what it meant.

But in what form? What material? What shape?

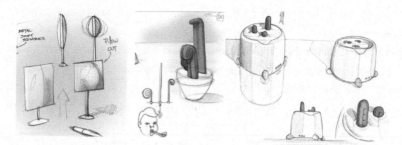

Sketches for shape-shifting, preattentive displays we explored at Ambient Devices.

Like the clock and the compass, the crystal ball has ancient roots. Various kinds of mystics and seers would gaze deeply into a cloudy-clear object or reflective surface—a gemstone, a rock crystal, a pool of water, a mirror—within which they could witness or conjure all kinds of things: past and future events, or seemingly unfathomable answers to questions. In *The Wizard of Oz*, for example, the Wicked Witch of the West tracks Dorothy's progress by gazing into her crystal ball.

In 2001 I reimagined the crystal ball as an orb—a grapefruit-size, frosted-glass oval, smaller than the ones that fortune-tellers use. The Ambient Orb picks up on the appealing shape and glow of the crystal ball, but fits on a desk, shelf, or counter and complements any style of decor. It's neutral in look and presence, without being bland.

The orb contains LEDs that change color in response to information coming in from a wireless connection. You choose the information source you want and the color range that appeals to you. For example, if you have allergies, you can connect the orb to the latest pollen counts. As colors change, you glance and know if the pollen count in the air is higher than usual. Sailors want to know if the wind is blowing just right for a sail. Gardeners want to be aware if there has been so little rain the garden needs some attention. Investors in the stock market want to track their portfolios. You can read the information in an instant, just like a clock, and not at all like a fussy alethiometer.

This solution worked. I could never have anticipated all its uses.

When venture investor Chris Sacca was at Google, he used the orb to show the height of the waves at a nearby surfing beach. His coworkers came to understand the color range, too. If somebody was looking for Chris, all it took was a quick glance at his green orb to know he would be gone for a few hours.

The orb has been adopted for use by companies in a huge variety of industries, and it's only one of many glanceable displays we will see in the future that will provide us with a glimpse of information, often all we need, in a beautiful, simple way—with color or movement or pattern or sound—rather than numbers and text, readouts and intrusive alarms.

PERVASIVE IS PERSUASIVE

Why do we want so much information? Why do we have this great urge to know what we cannot know directly from our senses? Sometimes we are driven by curiosity and the sheer delight of learning, but mostly we want to know more so we can make better decisions and take appropriate actions. Often making information ambient and pervasive is the most effective way to create behavior change for yourself and for others.

That was the idea behind the energy clock. Energy, like time and temperature, is invisible, except in its effects. We have wall sockets in every room, but are blind to the ramifications of plugging in another lamp or radio. In recent years we have finally started to understand that controlling our home energy use is essential to preserving our energy resources.

In the last ten years, at least twenty companies have tried to change their customers' behavior by developing different forms of home energy monitors to enable family members to see the invisible—how much energy they are consuming, and their overall load on the energy grid. When the load is high, energy *should* cost more. Setting the price of a service based on current demand is known as dynamic pricing. The purpose is to give people a financial incentive to use energy-consuming devices during low-rate times of the day (for example, running a dishwasher later at night). To do so, you need a device that will tell you what the energy usage is at any time.

The problem with previous attempts to create energy displays for consumer use is that the designs have been far from enchanted—downright boring. Some featured tiny, gray displays that were hard to read, and some spoke in a language that nobody understood, requiring users to learn the lexicon (e.g., kilowatt hours) of the electric company. Other displays translated usage into money, but the results were so unimpressive—"You saved 34 cents today!"—they didn't nudge people to change their behavior. Furthermore, information that arrived with the monthly bill wasn't actionable. It was hard to make the connection between the things you had done or hadn't done—like turning off the computer or leaving lights on all day—and an energy bill. The feedback loop was too long.

I wanted to find a way to make energy conservation compelling, so that people would be encouraged to change their behavior. The inspiration for what became the energy clock sprang from my rowing machine. The machine stores the data from my most recent workout so I can row against my own previous bests. I punch in the mode and two little boats pop onto the screen. One boat is me, today, and the other "pace boat" is also me, the last time I rowed. The effect is instantly motivating. I *should be able to beat myself.* There is nothing like competition to spur you on.

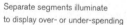

Separate segments illuminate
to display over- or under-spending

Under-spending
by 1 hour

Over-spending
by 3 hours

The energy clock motivates energy conservation by norming to yourself, in units you can understand. How many minutes are you ahead or behind your typical consumption at this time of day?

I took this idea of norming to yourself and translated it into home energy consumption. I chose the form of an augmented clock for the same reason Philip Pullman and J. K. Rowling did: people like clocks, are accustomed to them, and naturally glance at them throughout the day. No new behavior is required.

The display on the energy clock shows if you are ahead or behind your daily energy consumption and uses color to reinforce the information. A wedge of color appears next to the hour hand, representing your energy consumption for the day. Green means you're consuming less than you typically do. Red means you're an energy hog. The size of the wedge changes based on how many hours ahead or behind you are. No kilowatt hours or pennies are involved. The comparison to the "pace boat" of consumption motivates people to continually monitor and reduce their energy consumption, often through simple behaviors such as turning off lights and electronics when they're not in use and installing energy-saving appliances. The bar keeps changing, week by week, as their behavior changes. The pervasive potential of ambient displays, especially those in public spaces—such as offices, commercial buildings, airports, and schools—will help people learn more about and better monitor the health of their businesses, organizations, and teams, and even of entire cities.

This is precisely what has happened at SCE (Southern California Edison). In 2004, Mark Martinez, head of demand management of SCE, was on a long flight when he noticed an ad for the Ambient Orb in an in-flight magazine. He had been looking for a way to communicate with SCE consumers about the status of the electrical grid—letting them know when an extra-heavy load was on the system or when a brownout was imminent. The information was important and had to be delivered in real time, but there was no channel to do so. It didn't warrant an email blast, nor was it likely that customers would, of their own volition, go to a website to check on the status of their electricity provider. When Martinez saw the orb, he had a eureka moment. He thought it would be a perfect way to display the status of the power grid. It was simple, continuous, easy to understand, and cheap enough to distribute widely. He ordered several of them to test the idea. It worked. Now

SCE is distributing millions of the orb, renamed the Energy Joule, to customers throughout California. Businesses and home users can better plan their usage, and SCE is less likely to experience brownouts.

It may not be omniscience, as our fantasts have imagined it, but enchanted objects such as the Energy Joule come close: they enable us to see phenomena that we urgently need to know but have been hidden to us.

Beyond our desire to know all that can be known about the world around us lies the second and even more powerful urge—to know what is going on in the minds of other people: telepathy.

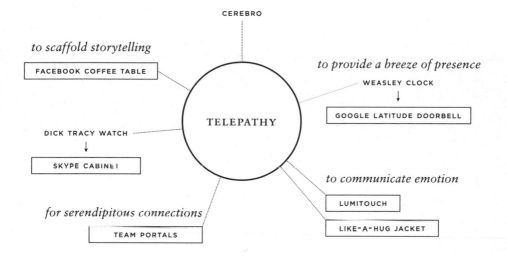

CEREBRO

to scaffold storytelling

FACEBOOK COFFEE TABLE

to provide a breeze of presence

WEASLEY CLOCK

GOOGLE LATITUDE DOORBELL

DICK TRACY WATCH

TELEPATHY

SKYPE CABINET

to communicate emotion

LUMITOUCH

for serendipitous connections

TEAM PORTALS

LIKE-A-HUG JACKET

TELEPATHY:
HUMAN-TO-HUMAN
CONNECTIONS

THE ACADEMY AWARD–winning movie *The Lives of Others* (2006) takes place in Communist East Berlin in 1986—the heyday of Cold War surveillance. The plot concerns an official of the Ministry for State Security, known as the Stasi, who orders one of his secret operatives to bug the apartment of a well-known playwright. What starts as political surveillance soon turns into romantic eavesdropping. The official and the playwright are in love with the same woman, and the official longs to know what she is doing and thinking when he is not with her. The movie is obsessed with the idea of surveillance—trying to know, through secret means, what is going on in the minds of others.

Although this desire is portrayed at an extreme level in the movie, attempting to connect deeply with others is a fundamental human drive. *What are people around me thinking? How do they feel about what's going on? What are they not revealing?*

FROM SPY WORLD

The drive to know what is going on inside other people's minds is a popular trope—from the mind readers and seers of ancient culture, to the Vulcan mind melds in *Star Trek*, to fantasies of Cold War spy and surveillance culture. More recently, in movies such as *Eternal Sunshine of the Spotless Mind*, *Inception*, and *Elysium*, pop culture has fixated on the brain and how it functions. In these movies, the contents of the brain are manipulated in various ways—memories deleted, mind states shared, and minds rebooted.

In the contemporary classic sci-fi movie *X-Men*, the character Xavier, the head of a school for mutants, can gaze into the minds of others thanks to a device called Cerebro, installed in the basement of the school. Xavier has a physical handicap and operates in a wheelchair—but Cerebro provides him with so much cognitive power that it more than compensates for his lack of mobility.

In the film, Xavier wheels himself into Cerebro, and once he's positioned, giant three-dimensional versions of the images in his mind appear around him. Not only can Cerebro project the contents of Xavier's thoughts, it can detect what's going on in the minds of the people around him, including his mutant students. This is telepathy: the power to read others' minds without using language or any other explicit method of communication.

A listening prosthetic from World War II used to track incoming planes before radar was invented.

In the real world, a number of technologies take a step toward telepathy. In the twentieth century, these were largely developed to aid in the gathering of hidden information. In 1921, before the invention of radar, the US Army created a mechanical listening device. The device consisted of two metal cones, each one twenty-five feet long and eight feet wide at the big ends, narrowing down to less than a foot in diameter. The cones looked like giant binoculars, but they acted like sound funnels, capturing incoming sound waves and funneling them into a pair of tubes the user placed in his ears. The effect was to dramatically amplify the user's hearing so that he could pick up the sound of enemy ships and planes that were still a great distance away. (A portable version of the device was created in Russia in the thirties.)[1] It wasn't telepathy, but it took a step in that direction by increasing the human ability to effectively listen in and divine the intentions (if not exactly read the thoughts) of another.

Listening devices and communication technologies were not confined to the armed forces. The inventor Alfred Gross helped bring radio technology, originally developed by the military, into general use. Gross invented the walkie-talkie, the CB (citizens band radio), the pager, and the cordless telephone. In 1946, the comic book artist Chester Gould met with Gross and was inspired to create the 2-Way Wrist Radio for his most popular comic book hero, Dick Tracy. The wrist radio instantly became a signature element of Tracy's character.

As we have so often seen, fictional objects inspire invention—another dialectic. Now we're seeing a slew of wrist-based devices coming on the market from companies such as Fossil, Microsoft, Casio, Pebble, Sony, and Samsung that behave remarkably like Dick Tracy's wrist radio. They're all trying to pack in as many features as they can—touch-screen displays, interchangeable faces, sensors, onboard storage, Bluetooth connection, USB ports, multiple apps—without getting too heavy or bulky and avoiding the need for too-frequent recharging. ("Parasitic" power—energy provided by light or motion rather than a battery or other embedded energy source—will eventually help with this issue.) The ultimate feature is the one Tracy got in 1964: 2-Way Wrist TV.[2] We may see it first from Apple—they have already registered a brand name for it: FaceTime.

In the 1970s and '80s, listening and snooping devices, both real and imagined, were the stock-in-trade of the spy culture. Growing up, I read and reread the James Bond books by Ian Fleming, including *Casino Royale, Moonraker, Diamonds Are Forever*, and *On Her Majesty's Secret Service*. I was fascinated by Bond's cars and gadgets, many of which were designed for gathering information that others wanted to keep secret. In *Octopussy*, 007 gives the evil Kamal Khan a fabulous Fabergé egg with a microphone and transmitter concealed inside, which enables Bond to listen as Khan reveals his nefarious plans.

Our fascination with ways of knowing continues. Plots of most spy-versus-spy movies of the current era—such as *Mission: Impossible* and the Bourne series—as well as 21 million cop shows, gangster films, and, more recently, terrorist dramas such as *Homeland*, hinge on the characters' ability to bug one another's apartments or offices and listen in on private conversations. No place is safe. Even when the characters try to lose themselves in public places, such as the parks of New York or Washington, DC, a van with a big parabolic "ear" is inevitably nearby, or a guy on a distant roof has a supersensitive microphone. Our drive to know what is going on in places we can't be extends even into sports. In football, the technician with the parabolic mike picks up the voice of the quarterback calling plays in the huddle or, in tennis, catches tennis players cursing themselves after dumping an easy volley into the net.

TELL-ALL CULTURE

Although the mechanisms of listening-in are still a popular media trope (how many more characters must be killed before they realize that wearing a wire is just too risky?), the irony is that the dominant cultural trend is not toward secrecy but toward openness and disclosure about what we're thinking, saying, and doing.

Today, we eagerly share information about our thoughts and actions—songs we listen to, videos we watch, articles we read—on Facebook, Twitter, Spotify, and other social media channels. This is a remarkable reversal of our surveillance-era view of control, and it represents a 180-degree turn of desire, a huge shift of agency. The telepathy-surveillance fear associated with the Cold War culture has morphed into a Facebook-era craving to know what our friends are thinking, from moment to moment, and a great drive for them to know our thoughts and feelings, too. We live today in a giant, invisible Cerebro known as social media.

Social media provides us with a kind of telepathy, a great deal of knowledge of others. We send and receive signals with our postings, photo sharing, tweeting, and texting. The problem is that we share far too many thoughts with far too many people, and a large percentage of them are "loose ties"—we have connected with them, but don't count them as close friends. As social networks grow larger and larger, and their traffic increases exponentially, we find ourselves living in a blizzard of text and images, and it is impossible to consume them as fast as they are produced. Even if you didn't need to work, eat, or sleep, you still wouldn't be able to drink everything that comes blasting through the fire hose of social media.

One of my students, Melissa Chow, created an enchanted jacket that inflates to give the wearer a hug when a friend likes their Facebook post.

We've gone from a quiet world where large ears are required to listen in to what others are thinking, to a cacophonous world where noise-canceling headphones are essential to our sanity. Filtering technologies and other methods of controlling the noise will become increasingly important, as will data visualizations that summarize information in ways we can absorb quickly.

Even with all this information available to us, the truth is that we don't really want telepathy of the Cerebro kind. We rarely want to know the explicit (unless you're a psychoanalyst) details of the stuff that's swirling around in the brains of those around us. Not only is there too much of it, but a great deal of it is irrelevant, distracting, and potentially disturbing.

What we *do* want is to have a better understanding of hidden thoughts and emotions that are relevant to us and would be important for us to know, and to be able to tune in to them—not through eavesdropping or mind invasion, but through a consensual, opt-in, mutually beneficial, benign telepathy.

Enchanted objects can play a role here by remapping social data onto objects that are a part of our daily routine and by enabling special kinds of awareness of people we care about.

Suppose, for example, that you had an enchanted wall in your kitchen that could display, through lines of colored light, the trends and patterns in your loved ones' moods? If you could understand that your spouse or child or parent had a regular pattern of emotions and begin to see how they connected to the environment, times of day, events, even your own moods, how would that change the relationships in your household? Would it make you a more attentive partner? A more effective caregiver? A more aware and understanding parent?

Would household members choose to share the status of their moods with each other? Why not? Not only do we have a difficult time reading the minds of others, we are not always good at expressing our thoughts directly and in the moment. In theory, we would like to better understand the moods and attitudes of those around us because they are not always clear or expressed. If technology could help us do that, without being intrusive or resulting in a version of the Heisenberg principle (which states, roughly, that an act of observation produces changes in what is observed), wouldn't that bring great benefit to our lives and our relationships? If we could know more about what's going on with those we love, we could alter our behavior in response. We might be quicker to celebrate the highs and good times of our lives together, more ready to offer support and understanding during low moments and difficult times. If we could see patterns of thought and mood in others, we might be better able to plan when and how we interact with them. In *Little Women*, Jo, the character most closely associated with the author herself, Louisa May Alcott, has a pillow that she positions on a parlor couch to indicate her current mood. *Leave me alone, I want to think.* Low-tech, but effective.

Enchanted technologies are just beginning to help with this kind of understanding. Today's objects aren't distracting and don't attempt to probe into deep emotions, but they do encourage connections and enable the surfacing of emotional content with people you care about. The Facebook coffee-table prototype, for example, anticipates the photos you might want to share with friends when talking about certain topics. The Skype cabinet makes conversation with a family member seem much more natural and human than sitting down in front of a computer screen. More on those later.

A PROKOFIEV-INSPIRED TELEPATHIC OBJECT

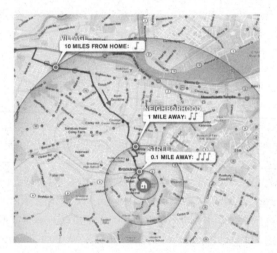

As each family member approaches home, the Google Latitude Doorbell plays a unique ringtone so people at home can anticipate their arrival.

The Google Latitude Doorbell lets you know where your family members are and when they are approaching home. When I was developing the Doorbell with Shaun Salzberg, a Media Lab student, we took as one of our inspirations the story of *Peter and the Wolf*. In this musical play, composed by Sergei Prokofiev in 1936, each of the main characters is linked to a unique melodic phrase, each played by a different, easily recognizable instrument. A clarinet signifies the cat, the wolf is a French horn, the bassoon announces the grandfather. When I was a kid, we used to listen to the record on my parents' giant console stereo.

At the Media Lab, as we were working on designs for the telepathic doorbell, the idea of associating a particular sound with a specific person came to mind. This idea fused with another inspiration: the Weasley family clock from J. K. Rowling's Harry Potter series, which is also a kind of telepathic device. It helps the Weasleys distill and summarize information, rather than overwhelming them with more than they need. All the clock does is keep track of their children—Ginny, Ron, Fred, George, Percy, Charlie, Bill, Molly, and Arthur. Like the signifying tunes in *Peter and the Wolf*, each of the nine hands of the clock is associated with a member of the family, and each hand is marked with that child's name and picture. The ornate dial is inscribed not with hours, but with words: *home, hospital, prison, lost,* and *mortal peril.* As Rowling writes, the clock "is completely useless if you want to know the time, but otherwise very informative."[3]

The brilliance of the Weasley clock design is the way in which it prioritizes and displays information. Rather than showing the family members' physical locations on a map, it provides something much more important: knowledge about where they are, in general, as well as their state of being. *Lost* is shown as the opposite state to *traveling,* although the GPS position would appear the same on Apple's Find My Friends app map-based display.

The Google Latitude Doorbell, a prototype I built at the Media Lab, plays a similar role in family life. In particular, it helps with those sometimes frustrating, and even fraught, transitions at the end of the day. One family member is trying to plan dinner so that everyone can eat together. Another one is stuck in a meeting and can't communicate what's going on. A third is on the playing field and the game is running long. Another is stuck in traffic. The Google Latitude Doorbell, combining inspiration from Prokofiev and J. K. Rowling, represents each person you want to keep track of by a distinctive chime. As each family member approaches home, the chime sounds for that person when he or she is ten miles away, one mile away, or a tenth of a mile away . . . nearly home.

It's not telepathy, but it does deliver information that gives clues to the mental and emotional state of each person. Frustration for the unlucky one in the traffic jam. Exhaustion, with possible elation or crestfallenness, for the athlete. Mental distraction from the person in the intense meeting. The chiming also helps the person preparing dinner get the timing right, so the meal will be ready when the whole family is assembled to share it.

THE NEAR-TELEPATHY OF MUTUAL PRESENCE

Telepathy is communication with other people without the involvement of any of the five senses—we simply *know* what they are thinking and feeling, without having to do or say anything. As I've said, technology is getting us closer to that ideal by offering the ability—through Twitter and IM conversations, Facebook walls, and other such methods—to have a continuous, if thin, awareness of others who are close to us or not close at all.

THE VALUE OF PRESENCE

Enchanted Objects support a new genre of continuous, lightweight awareness.

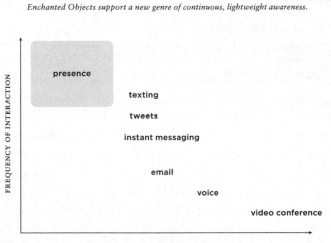

The key is mutual presence. Even now, when physically together, we pay continuous, if partial, attention to one another, through low-involvement conversations and interactions. This constant consensual connection will become even more of a regular feature of our lives, but less obvious, with smaller and smaller Bluetooth headsets and always-on connections. You could be in your workspace with remote colleagues visible through Skype on a screen nearby and your assistant connected by a Bluetooth headset.

A nice example of lightweight mutual—also known as reciprocal—presence is the LumiTouch picture frame, developed at the Media Lab by Angela Chang, Ben Resner, Brad Koerner, XingChen Wang, and Hiroshi Ishii. A framed photograph is a central keepsake for people. It often stands in for a personal connection. It reminds us of moments in time and the emotions and experiences associated with them. We look at the people in the image and wonder about them. *How are they? Do they think of me as I think of them? I wish I could be in touch.*

LumiTouch frames come in pairs and glow when the other person approaches their frame. Squeeze the edge of the frame to illuminate that edge and "blow a kiss."

The LumiTouch frame concept picks up on this emotional connection and enables people who are physically separated to detect each other's presence and communicate feelings—in a lightweight fashion. You and your friend each have a LumiTouch frame at your location, and the two frames are connected over the Internet. When you squeeze the frame, your friend's frame lights up. The color shifts in response to how hard and how long you grip the frame. Your friend can simply enjoy the sentiment represented by the shifting colors on the frame or reciprocate by squeezing back.[4] What's pleasing about LumiTouch is that it enables communication without requiring active and focused participation. However, the squeezing may also serve as a signal that it's time for a more active connection—a text or a call.

Much has been said about the human *disconnection* that comes from the connected society. Sherry Turkle, in particular, has been thinking for years about how technology affects human relationships. In her book *Alone Together* she writes, "Technology is seductive when what it offers meets our human vulnerabilities. And as it turns out, we are very vulnerable indeed. We are lonely but fearful of intimacy. Digital connections and the sociable robot may offer the illusion of companionship without the demands of friendship. Our networked life allows us to hide from each other, even as we are tethered to each other. We'd rather text than talk."[5]

I believe that enchanted objects—with their persistent presence and lightweight background communication—will have a different and positive effect. We will be more connected to the mood states, behavioral changes, and communication styles of those we live and work with, the people we care about, think about, and love. We will be able to customize, personalize, and tailor our communications and connections through these technologies. We will take a step closer to one another rather than be pushed farther apart.

to protect without harm

STAR TREK PHASER ⟶ TASER

for invisibility, to avoid harm

POTTER'S CLOAK

SAFEKEEPING

being prepared

FRODO'S SWORD STING
↓
AMBIENT UMBRELLA

to guard and track

LOCKITRON LOCK

TILE

TAGG DOG COLLAR

for accountability

CAR CAM

GOOGLE GLASS

SAFEKEEPING: PROTECTION FROM ALL HARM

IN SEPTEMBER 1966, at the height of the Vietnam War, *Star Trek*, the twenty-third-century space adventure, premiered on NBC and ran for three seasons, then spawned four more television versions and a string of incredibly popular movies. Gene Roddenberry, the creator, was inspired by old Westerns and Jonathan Swift's *Gulliver's Travels*. But he was also a creature of his times, affected by the growing student and youth culture that protested the Vietnam War, and hoped for a better, more peaceful future. Roddenberry said that by imagining a whole "new world with new rules, I could make statements about sex, religion, Vietnam, politics, and intercontinental missiles." Roddenberry was delivering his messages to millions of viewers and, as he put it, "fortunately they all got by the network"[1] executives who might have censored them if the show had been a realistic drama.

Not only did *Star Trek* explore the cultural themes of the day, transposed into twenty-third-century action, Roddenberry imagined a variety of technologies that aligned with those themes. Replace outer space with Vietnam, and you can see that Roddenberry was looking for a way that people from one society could explore the world and interact with people in other societies without invading or harming them, but also keeping themselves protected from harm. From this hope came the phaser and its ability to neutralize people without destroying them. In the *Star Trek* episode "A Piece of the Action," a group of tommy-gun-wielding mobsters, shooting at each other in the street, all fall to the ground when blasted by the *Enterprise*'s phaser, which Captain Kirk has commanded be set to stun. "Set phasers to stun" has become a cultural meme.

To be safe, to feel protected from harm, is one of our deepest, most basic human drives, and over the centuries much of our tech innovation has focused on this drive. We have invented devices—in our imaginations and in reality—that would keep us from all kinds of harm and anxiety. Natural disasters. Illness. Pestilence. Evildoers. Many of these devices were first created for combat and were intended to harm others before harm could be done to us, but we have also wanted to create protection technologies that would neutralize a threat, keep us safe, without resorting to violence or causing damage to others.

Enchanted objects, including those of safekeeping, have typically responded to a specific cultural need. We invent or popularize characters, create and elaborate stories, and imagine or exaggerate the importance of technologies that align with the zeitgeist—the spirit of the times—just as the phaser was an antidote to the pain we felt in a Vietnam-era society that was being torn apart by the use of killing technologies. The tales that swirl around folkloric characters such as Paul Bunyan, the giant lumberjack, and real but mythologized pioneers: Daniel Boone (1734–1820), Davy Crockett (1786–1836), and Johnny Appleseed (1774–1845). Bunyan had his enormous ax, to harvest trees and clear territory for homes and fields to farm, as America expanded west. Crockett had magical skill with his long rifle—able to fire a bullet from forty yards and hit the edge of an ax, splitting the bullet in two—a handy frontier skill to ward off marauders and bears and find your next meal. Davy Crockett, it was said, could "wade the Mississippi, leap the Ohio, [and] ride upon a streak of lightning."[2]

NONLETHAL WEAPONS OF SAFEKEEPING

In the mid-1960s, while we were witnessing horrific scenes of real-life warfare on the news, in *Star Trek* the crew of Starship *Enterprise* was handling conflicts with the phaser. Unlike the M16s and armored tanks of the Asian land war, Captain Kirk and Mr. Spock and their crews could deflect lethal violence with a simple neutralizing blast of the stun gun.

I had assumed that the Taser, the real-life stun-producing weapon now used by police around the world, was a realization of the *Star Trek* phaser, but it actually has a different provenance. The Taser was developed by Jack Cover, starting in the late 1960s. His inspiration was Tom Swift, the protagonist of dozens of novels, the first of which, *Tom Swift and His Motor Cycle*, came out in 1910. Just as the tensions of the Vietnam era gave rise to *Star Trek*, the early-twentieth-century excitement of inventions and technology—with Thomas Edison as the iconic genius-tinkerer—brought forth Tom Swift, an action-loving boy who was fascinated by science and technology and was constantly conjuring fantastic technologies and new adventures in which to use them.

One of Tom's inventions presaged the *Star Trek* phaser. In *Tom Swift and His Electric Rifle*, Tom displays the new device to his friend Ned, who asks how it works. "By means of a concentrated charge of electricity which is shot from the barrel with great force," Tom replies. "You can't see it, yet it is there."[3]

Taser, according to Cover, is an acronym for *Tom Swift's electric rifle*. The *a* was added only because Cover got "tired of answering his phone TSER."[4] The Taser continues the "set to stun" model of the *Star Trek* phaser, using neuromuscular incapacitation (NMI)—a less alarming term for a massive electric shock—which, according to the Taser company, "temporarily overrides the command and control systems of the body to impair muscular control."[5] The company markets its wares to law enforcement, the military, and consumers and describes them as self-defense products that reduce violence and save lives. On the Taser website, a digital counter keeps track of the "total number of lives saved" by the self-defense device, which, when I checked, was over one hundred thousand. The Taser also has some memory; it keeps track of every action a user takes with the device (including when it is shot), and the data can be transferred to a computer through a USB connection—which, soon enough, will likely be replaced by a wireless connection to the cloud, then geo-coded so we can all see Taser hot spots in real time.

BENIGN PROTECTION?

Rather than focusing on technologies of incapacitation, let's move to more mature and advanced forms of safekeeping—the life-affirming kind in which you keep yourself safe but not at the cost of others.

One form of safekeeping is accountability—applying technology to alter the behavior of people in situations that could be hazardous to others. England has deployed some 2 million "eyes" in public spaces, or about one camera for every thirty-two citizens.[6] These cameras began to proliferate in the 1970s, in part as a way to deal with the threat of terrorism.[7] Their use is now justified as a deterrent to crime, which I doubt it is, and as an aid in solving crimes after they have been committed, which I have no doubt is valuable.

In addition to this massive increase in the number of surveillance eyes, we have a huge proliferation of cameras in smartphones and tablets and other devices. There are cameras everywhere. On the sunny spring afternoon of April 15, 2013, I was watching the Boston Marathon from a good vantage point on Beacon Street when the bombs exploded. Within hours of the terrorist blast, authorities—including investigators from the FBI, Homeland Security, and the Boston Police Department—began collecting and analyzing evidence. Some of their most important sources of information were the images gathered by the many surveillance cameras mounted in buildings and public spaces throughout the city, those captured by people watching the race, and by television crews. Soon, law enforcement officials announced they had found footage of a suspect placing what appeared to be a backpack containing an explosive device at the scene of one of the explosions. That day, no one protested that the surveillance network was a threat to our civil liberties. Quite the opposite. This crowdsourcing of criminal investigations will continue to grow in effectiveness as more citizens wear more cameras in clothing and on their glasses. (The Boston Police Department runs a Crime Stoppers program—established to gather tips from the general public about unsolved crimes—which is the name of Dick Tracy's ad hoc group of crime fighters.)

Taken together, the de facto surveillance network goes a long way toward fulfilling one of the fundamental human desires—safety—with a strong dose of omniscience to boot. I'm not worried about the uses of surveillance technologies in countries with strong constitutional protection of the right to assemble and to free speech. When those safeguards are in place, surveillance technologies offer *more* opportunities for safekeeping, as in the Boston Marathon bombings investigation, and they outweigh the risks of unwarranted and damaging intrusion.

Another story of how surveillance technology can protect us through accountability comes surprisingly from Russia, the land of the KGB: the dashcam. The dashcam is exactly what it sounds like it is—a digital camera mounted on the dashboard of your car, with the lens directed at the road ahead. Although dashcams are ubiquitous in Russia, I didn't pick up on the trend until a meteor seared a trail across the sky over the city of Chelyabinsk in February of 2013. Because thousands of drivers were on the road with dashcams pointed in the general direction of the meteor's path, that amazing event was caught in a deluge of footage.

Safekeeping, not the capture of random events, is the primary purpose of the dashcam. Driving in Russia is dangerous. According to the World Health Organization (WHO), driving accidents caused 26,567 fatalities in Russia in 2010,[8] or about 18.6 per 100,000 citizens.[9] This is much worse than the rate in the United States (11.4), but nowhere near as bad as in some other countries, such as the Dominican Republic (41.7).[10] Dashcams, however, aren't primarily for accidents but to help Russians deal with the authorities. Traffic cops are notoriously corrupt. When violations or accidents take place, claims are made, and court cases result, but people do not trust the police to provide faithful accounts of what happened. Dashcam footage has become an essential source of evidence. The dashcams are, in one way, voluntary and opt-in—people buy their own and mount them in their cars as they choose. But they are not so very different from the surveillance cameras in the United Kingdom in that I have no choice about being captured by your dashcam if I drive past you. Russians have accepted dashcams as a fact of modern life. The Russians' drive for safekeeping, coupled with the high risk of harm, is stronger than the worry about loss of privacy.

Dogs, babies, and our elderly are made to wear tracking tech for safety's sake. This onesie by Mimo measures your infant's respiration, skin temperature, body position, and activity level.

If you think of the use of observance technologies as monitoring, rather than spying, it takes on a very different meaning. A baby monitor gives parents a sense they are doing all they can to protect their child and the well-being of their family. The GPS device on your dog's collar enables you to find her if she runs astray or to keep tabs on the dog walker when you're away.

I have personal experience with living under surveillance every day, and I can attest that I find it far more reassuring than discomfiting. For twelve years, I have kept a small office at the Cambridge Innovation Center—a twelve-story building owned and managed by MIT. It is home to independent entrepreneurs like me, start-up companies, and some more established ones. Google, in its early days, had an office there, as did Apple and Amazon. It is a wonderful place to work for many reasons. You rub shoulders with fascinating people at the little self-serve cafés that grace every floor. You have high-speed Wi-Fi in every nook and cranny. The offices and conferences have an academic feel about them—well designed and well equipped without anything unnecessary or over-the-top like the ones you might find in a corporate facility.

The feature that is relevant to the discussion of safekeeping is the security cameras that are ubiquitous throughout the building. Every six feet or so, along every hallway, in every office and conference space, is a camera. Although I was vaguely aware of them when I first moved in, I quickly forgot about them and gradually came to see them as valuable. Why? Because everyone feels safe, protected, watched over. When you go to lunch, you can leave your laptop on your desk and not bother to lock your door or even close it. About once a year something disappears from where it is supposed to be. But it is always recovered because every inch of the building and every second of the action has been recorded. The security people can send out an image of the missing object or of the person who snitched the object to everyone in the building. Soon enough, the purloined or misplaced object is returned or the culprit is identified.

After Ambient Devices developed the Ambient Orb, we installed five of them at the entrance to our shared office space. Each orb displayed weather in one of five cities around the world. One day one of my officemates reported to the concierge (who looks after things) that an orb had disappeared. A quick review of the video revealed who the culprit was: me. I had swiped the orb to use in an investor meeting. Within an hour I received a polite email from the concierge asking me to please return the object. I did.

Security at the CIC is wikilike—everyone contributes. Partly because of this ability to tap into our telepathic security system, if and when needed, we all know we are accountable to one another. That makes us feel protected and also feel responsibility to be protective of others.

THE PROBLEM
OF BORDER CROSSINGS

Using cloud-connected objects for protection demands new policies and laws to ensure our privacy. Ben Franklin famously said, "Those who would give up essential liberty to obtain temporary safety deserve neither liberty nor safety."[11]

In 2010, the Indian government announced plans for a universal ID system, called Aadhaar, which would collect biometric data—including fingerprints, an iris scan, and a picture of the face—on every one of the country's 1.2 billion people. In addition, the system would collect routine data, including name, gender, date of birth, and address. Each person would receive a card that would serve as identification for all the purposes one might expect, including access to bank accounts and as a payment card.

One of Aadhaar's major purposes was to better administer social welfare programs. Some 300 million people in India have no identification or vital records. Without knowing who is in need and where to find them, it is difficult for the government to deliver aid and support in any organized way. And without being able to prove who you say you are with authentic documentation, it can be hard to collect benefits that are due you. Despite the difficulties involved in gathering information about more than a billion people (one of which is that many manual laborers have worn away their fingerprints through hard work), the government is plowing ahead with the effort.[12]

Although the government's stated goal is a form of safekeeping, it is easy to understand how people might feel otherwise. They have possession of my biomarkers! What if corrupt government officials hack the database? What might ensue? Could they steal my identity? Empty my bank account? Do I want to opt into a government-controlled safekeeping system?

We are good at imagining the worst and expressing it in our fictions because it is so effective at driving a compelling narrative. In *The Dark Knight*, Batman deploys the superadvanced technology of Wayne Enterprises to connect the cell phones of all the citizens of the great city of Gotham into a massive surveillance system with the goal of capturing his nemesis, the Joker. Gothamites are given no choice about opting in. Batman, a public servant who has devoted his life to the safekeeping of the people of his beloved city, is quite willing to invade a little privacy to gain an advantage on evil. In the movie *Gattaca*, human beings are genetically engineered and the state limits the freedoms of those who are in some way imperfect.

Our fears are well-founded. A friend with long experience in the automotive industry told me that today's Internet-connected systems are linked to control systems, and by hacking them, a terrorist could "drive every car off the road in an instant." Luckily, those networks are closed loop and not connected to the Wild West of the open Internet.

Adam Greenfield, author of *Everyware*, describes how technologies developed with positive intentions can be exploited in unexpected and sinister ways. Greenfield offers the example of a business implementing a radio frequency ID (RFID) card system, which can store a "lengthy list of attributes about the person carrying it." The state government notices the card and decides to adopt the private scheme for public use, requiring its citizens to carry one, just as the government of India is doing. The tricky part comes when other "private parties" realize the usefulness of the ID cards. All they need is a "standard, off-the-shelf reader" and they have the "ability to detect such attributes, and program other, interlinked systems to respond to them." For example, the management of a private club might decide to bar admittance to people with certain characteristics. Greenfield concludes, "No single choice in this chain, until the very last, was made with anything but the proverbial good intentions. The logic of each seemed reasonable, even unassailable at the time it was made. But the clear result is that now the world has been provisioned with a system capable of the worst sort of discriminatory exclusion."[13]

MIT professor Gary T. Marx argues that the use of surveillance tools and universal ID systems and other technologies for gathering and storing personal information increases the potential for "border crossings"—which is what happens when information that is intended for one use finds its way into a different domain, with possible deleterious effects.[14]

The headline of a *Forbes* article succinctly describes just such a border crossing: "How Target Figured Out a Teen Girl Was Pregnant Before Her Father Did." As with most retailers, Target records information about purchases then analyzes the data to devise sales and promotional offers for different categories of shoppers.

In this case, Target understood that pregnant women buy certain items—such as lotions and supplements—at well-defined stages of their pregnancy. When Target determined that a young woman was following this buying pattern, they sent some coupons for relevant items to her home. When the father of this teenager collected the mail, he found the pregnancy offers addressed to his daughter. He was shocked. His first impulse was to complain vigorously to Target, but he had to apologize when he discovered that his daughter was indeed pregnant.[15] Target had crossed the border that (barely) separates consumerism from family life or, more simply, from public identity to private persona.

Marc Langheinrich, an assistant professor of computer science at the Università della Svizzera Italiana and head of the Research Group for Ubiquitous Computing, agrees with Adam Greenfield about the privacy issue created by connected things. He believes that one of the most important concerns regarding ubiquitous computing and its effect on privacy is how infrastructure can be used to support search techniques and how those techniques will result in border crossings. "Ubiquitous computing systems, even when installed for the greater good and with the best of intentions, will run a high chance of involuntarily threatening our personal borders that set apart private from public."[16]

How can we protect ourselves from systems that are designed to protect us, but don't? I believe the most obvious safeguard in the Internet of Things is transparency. Consumers should always be given notice when information is being collected and should be able to give consent or to opt out. They should also have access to the data that has been collected and have assurances that the data is secure.[17] (It doesn't mention anything about what to do when your father receives a coupon for baby merchandise in the mail.)

OBJECTS THAT ANTICIPATE THEIR USE; KNOW WHEN THEY'RE NEEDED

Safekeeping can come in much smaller and less complex enchanted packages than buildingwide security systems or countrywide ID programs.

The company I cofounded, Ambient Devices, thinks about safekeeping in a much more personal and everyday application—being prepared for the weather. No, rain and snow are not as life-threatening as the approach of orcs, but every family has to cope with the weather every day. The Ambient umbrella has one enchanted feature—to prompt you to take it with you when you head out the door. A wireless receiver in the handle of the umbrella connects to the nationwide Ambient network and receives data from AccuWeather for your zip code. If rain is forecast, a ring of LEDs embedded at the top of the umbrella's handle glows and pulses a gentle blue light. After owning the umbrella for a while, you come to expect and appreciate its modest but valuable contribution to your safekeeping.

The door lock is another of our oldest and most basic devices of safekeeping. Although it's easy enough to use, the interaction is rarely delightful. There is the fishing of the key from the purse or pocket, the constant worry of losing it, and that moment of anxiety as you fit the key to the lock, eyes fixed on your task, when you feel vulnerable—might an intruder rush in from behind?

Lockitron created a Wi-Fi–connected device that fits on an existing door lock and enables you to control it and know if it is locked. It will automatically ensure that you lock the door when you rush out each morning. It will inform you if a member of the family has locked or unlocked the door. It enables you to control the lock remotely, to allow in a member of your family who has forgotten the key or admit a plumber when no one's around. People who list their apartments on Airbnb, the worldwide accommodations rental service, use the device to give access to their guests without having to be at home when the renter arrives. Lockitron gives a feeling of protection and safekeeping and obviates the need for a key. A pulse from your smartphone as you approach the house opens the lock.

Another form of safekeeping is to never be lost. That's the function of these little wireless tags from a start-up called Tile. Now your phone can find your keys or whatever else you tag.

Lockitron and the Ambient umbrella are two of the interface-enchanted objects that link an existing, unsmart piece of infrastructure with the cloud—they are modest bespoke devices, simple to understand and use, but ones that produce large effects: the feeling that we are prepared and shielded from the elements and our safety is ensured.

Just beyond safekeeping lies another ancient and fundamental human desire: the wish for a long, healthy, protected life—and even immortality.

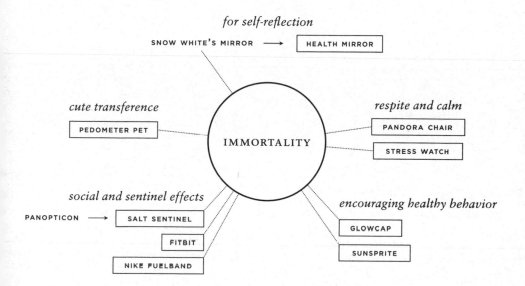

for self-reflection

SNOW WHITE'S MIRROR ⟶ HEALTH MIRROR

cute transference

PEDOMETER PET

respite and calm

PANDORA CHAIR

STRESS WATCH

IMMORTALITY

social and sentinel effects

PANOPTICON ⟶ SALT SENTINEL

FITBIT

NIKE FUELBAND

encouraging healthy behavior

GLOWCAP

SUNSPRITE

IMMORTALITY: A LONG AND QUANTIFIED LIFE

IF ONLY THERE were truly a potion, an object, a place, a magical compound, that would grant us health, strength, healing, and vitality. From the earliest myths to the most recent video game, this desire for immortality has manifested itself in myriad tales and inventions. The philosopher's stone. The elixir of life. Fountains of youth. Pharmaceuticals such as the Crazy Alchemist's Potion, lost somewhere in the *World of Warcraft* game.

In our immortality wish, just as with the other fundamental desires, we find dialectic pairs of objects, fictional and real, reaching back to earliest times. In the *Epic of Gilgamesh*, one of the oldest recorded stories (written on a tablet, albeit a stone slab rather than a glass one—actually twelve of them), a king in search of the elixir of life embarks on a journey to a distant land where two immortals live. It turns out that a water plant contains the secret substance. The king plunges into the ocean, grasps the plant, only to have it plucked away by a snake.

Herodotus, the Greek historian writing in the fifth century BC, tells a story of Ethiopians who live to be 120 years old. Their secret seems to be a spring where they regularly bathe and become "sleek of skin, as if it were a spring of oil; and from it there came a scent as it were of violets."[1]

The health-giving and restorative power of water is a common theme in health fantasies. People still head for Florida in search of the modern equivalent of Ponce de León's fountain of youth. In Ron Howard's movie *Cocoon*, elderly Floridians grow younger as they bask in a pool at their nursing home.

Health-giving cures and potions are often concocted from magical ingredients. One of my favorites comes from *The Chronicles of Narnia*, by C. S. Lewis. Remember Queen Lucy the Valiant? Her vial, shaped from a diamond, contains a cordial made from the juice of fire flowers. It can cure illnesses, heal wounds, and revive those on the brink of death. The berries of the fire flower, too, have extraordinary power—they can make the stars grow young.

Potions are just as popular in the fictions of today. In *Indiana Jones and the Last Crusade*, Indy pours healing water from the Holy Grail onto his father's surely mortal wound. The water fizzes and fumes as if thrown onto a fire's embers, the wound closes up, and Dr. Henry Jones is restored to life. Characters in *World of Warcraft*, the most popular multiplayer online video game, can ingest a remarkably long list of potions and elixirs to make themselves stronger, faster, or enhanced in numerous ways. Potions include Insane Strength, Indestructible, Swiftness, Swim Speed, Restorative, Curing, Healing, Rejuvenation.

Enchanted objects, too, can impart magical abilities to humans and promise long life. In the short story "The Philosopher's Stone," written in 1789 by Christoph Martin Wieland, the character King Mark of Great Britain falls under the spell of Misfragmutosiris of Egypt, who claims he knows how to concoct the philosopher's stone. However, it won't be easy to do and it won't come cheap. To put together a "true hermetical philosopher's stone," says the Egyptian, you need to assemble "the finest precious stones: diamonds, emeralds, rubies, sapphires, and opals" and then mix up a "large portion of red mercury sulfide and some drops of oil drawn from a condensed ray of the sun." The whole operation takes about three weeks and results in a "kind of crimson substance, which is very heavy and can be scraped into a fine powder, of which a pinch half the size of a barley seed is sufficient to transform two pounds of lead into just as much gold."[2] Needless to say, this is a hoax designed to drain the king of his resources. One moral of the story, which applies today, is that price influences the perception of value. The more we pay for our medications, the more we believe they are doing us the good we desire. The placebo effect is increased and the potion is more effective.

The Harry Potter books owe part of their incredible appeal and popularity to Rowling's supple updating and remixing of the fundamental human drives in ageless objects of enchantment and magic. The list of enchanted objects that animate the series includes the philosopher's stone, as well as concealing devices, detecting objects, mirrors, transporters, and other forms of magic that require more or less dexterity or training to deploy. None of them is quite so complex, however, as Pullman's alethiometer.

In the final book of the series, *Harry Potter and the Deathly Hallows*, we learn of three objects, the "hallows," that provide power to a sorcerer: the elder wand, the resurrection stone, and the cloak of invisibility. All of them bring some power to ward off death. With the wand, you cannot be defeated in battle. With the resurrection stone, you can communicate with the departed. With the cloak, you can make yourself invisible to everyone—including death itself. Our fear of mortality drives these narratives forward.

Inevitably, these attempts to discover a source of perfect health and immortality fail. They often lead, however, to a revelation for the seeker: to make the most of life while we have it and seek immortality in other ways, such as our faith or our legacies.

As Gilgamesh is advised by a wise barmaid:

> *The life that you seek you never will find:*
> *when the gods created mankind,*
> *death they dispensed to mankind,*
> *life they kept for themselves.*

But you, Gilgamesh, let your belly be full,
enjoy yourself always by day and by night!
Make merry each day, dance and play day and night![3]

THE SINGULARITY AND TRANSHUMANISM

As much as we may take this message to heart, the fundamental human drive for perfect health and longevity is as powerful as ever. Humans today are no less prone to believe in potions and elixirs, magic stones, fountains, and youth-giving springs than our ancestors. We now call them medicines and treatments, surgeries and preventions, diet and wellness regimens.

Technology plays a huge role in the fulfillment of this drive. Ray Kurzweil, one of the most visible advocates of long life, the "singularity," and transhumanism (both deal with the eventual fusion of man and machine in everlasting, cyborglike beings), is working on a variety of ways to combat disease and prolong life. He predicts that, within fifteen years, life expectancy will be increasing by one year every year. This will largely be accomplished by steadily eliminating the most common causes of death, such as heart disease and diabetes. (Kurzweil suffers from glucose intolerance and reportedly takes some 150 supplements a day to control it.) Kurzweil predicts that by 2030 or so "we'll be putting millions of tiny, single-purpose robots called nanobots inside our bodies to augment our immune system and wipe out disease. One scientist has already cured type 1 diabetes in rats with a blood-cell-size device."[4] By 2050, Kurzweil says, our entire body might be composed of nanobots and we will be completely disease-free. That may not be eternal life, but it ensures a longevity well beyond our current calculation of it.

In the quest for immortality, fantasy and the technology overlap in odd ways. Did Michael Jackson really have an oxygen chamber in which he slept to boost his capacities? A man called FM-2030 (born

Fereidoun M. Esfandiary) was a transhumanist, author, professor, and former basketball player. Like Kurzweil, he envisioned a new world in which technology and its spread would fundamentally alter our basic human functionings and assumptions. He hoped to live to be at least a hundred, which would have meant celebrating his centenary birthday in 2030. "The name 2030 reflects my conviction that the years around 2030 will be a magical time," FM-2030 said in an interview on NPR. "In 2030 we will be ageless and everyone will have an excellent chance to live forever. Twenty thirty is a dream and a goal."[5] He died in 2000, from cancer, and now resides in cryogenic suspension at a facility in Arizona.[6]

Most of us do not genuinely yearn for immortality. We've seen how hard it is for vampires to cope, at least after those first couple hundred years of constant fun. Philosopher Stephen Cave, author of *Immortality*, talks about four paths to prolonging life: Staying Alive, Resurrection, Soul, and Legacy. The first of these—staying alive—is where we put a good deal of our daily energy today, and it is the one in which technology and enchanted objects most come into play. Cove argues that the promise of defeating disease and debilitation has "never been more widespread than today. . . . A host of well-credentialed scientists and technologists believe that longevity liftoff is imminent."[7]

I am less concerned with "longevity liftoff" than I am with finding ways for enchanted objects to help us achieve maximum well-being by taking advantage of the health boons already available to us.

Beam rewards kids with prizes for regular brushing. Do we really need to quantify every daily activity?

THE QUANTIFIED SELF

Like most Americans, I've been trying to lose fifteen pounds for about twenty-five years, and I have tried all kinds of motivational stunts—such as signing up with a few friends to compete every year in a June triathlon on Cape Cod. To help train, I have been working with a fitness coach for several years. We have a good relationship. Last year, after a few weeks working together, she assigned me to do a deceptively simple task: "Every day, I want you to keep track of everything you eat. Write it down in this journal." She handed me a small, black Moleskine notebook. "And I mean everything. I'm *not* asking you to *change* anything about your eating habits. Just write it all down."

I had no idea what a powerful effect this task would have. As I started to keep track of everything that entered my mouth, I found things I preferred not to write down. Yes, I wanted to complete the assignment honestly, and I had no intention of deceiving my trainer, but I also did not want to record on paper the three oatmeal cookies I usually gobble down in the afternoon or the chewy peanut-butter granola bars I like to snack on.

So, although my trainer hadn't asked me to change anything about my eating habits, I did moderate my diet. I found it was almost as pleasing to eat an apple and a pear for a snack instead of my regular ration of cookies, and it also looked so much better in the notebook. I wanted to be honest in my reporting, to prepare properly for the triathlon, and to please my trainer. The social incentive was essential. That, combined with the quantification, led to a change in my eating behavior.

However, I couldn't sustain the effort over time. I didn't want to carry the notebook with me wherever I went. Recording my consumption was tedious and inconvenient. I needed an enchanted object to help me, not a stubborn notebook. As we have seen, an entire world of connected sensors can be put to work in such efforts. Our shoes, clothing, cutlery, even our salt shakers, can be enchanted. They enable us to hold up a new kind of mirror to ourselves, to better witness and analyze the decisions and actions we take every day. If needed, these devices can share the information we collect with other people—a trainer, doctor, coach, adviser, or that set of triathlon friends—who can aid us in meeting the goals we set for ourselves, whatever they are.

What kind of enchanted object could have done the work of the Moleskine to help me keep track of what I ate and make it available to others? I have plenty of product concepts in mind. Some of them would be relatively easy to implement and deploy, while others—such as wireless jewelry—would require a different kind of discipline to create.

Here are the essential elements of my enchanted-object recipe:

- An ordinary familiar object, augmented and connected to the cloud.
- Passive sensing, so you don't have to manually record anything.
- Unavoidable, ambient information display for constant feedback.
- Emotional engagement and/or social incentives.

And here are a few ideas for enchanted objects in service of the quantified self:

MyFoodPhone. This is a service I helped develop for Sprint at a workshop at Canyon Ranch in the Berkshires in 2002. Before you eat, you use your phone camera to take a shot of the plate of food or the snack you plan to consume, and the service sends the photo to a dietitian, who evaluates the food and sends back a summary of the calories, grams of carbs, protein, and fat. (Today, image recognition could replace the live dietitian.) MyFoodPhone was marketed by Sprint a few years later as a subscription service, at $49 per month. A like-minded smartphone app, The Eatery, offers similar service for free since the information is crowdsourced. The analysis is less precise, but evaluation of the photos offers a social aspect that MyFoodPhone doesn't. You gain the motivation from knowing that hundreds of people are offering their opinions on every meal you eat, and that you will have the opportunity to return the favor by evaluating and commenting on their food choices.

Tattletale jewelry. A Japanese researcher whose specialty is obesity visited MIT and showed us a working prototype of a necklace with an embedded microphone and microprocessor. You wear the necklace throughout the day, and whenever you chew, the distinctive sound of mastication is picked up by the microphone and delivered to your phone so you receive feedback through text messages. When your snacking trends high, you may well be motivated to cut back.

HAPIfork. The HAPIfork was introduced at the Consumer Electronics Show in 2013 and got a lot of attention, both as an innovation and also as a response to the obesity crisis in this country. "Buzzing Fork Offers Ultimate First-World Solution to Overeating," read the headline in the *Huffington Post*.[8] The goal of the HAPIfork is to record your food consumption and help you eat more slowly and intentionally and, therefore, less. The fork contains a sensor that keeps track of each time the fork is used. If you're eating too fast, an indicator light flashes, signaling it's time to take a break. The fork records the total duration of your eating, the number of "fork servings" you take per minute, and the time that elapses between each time the fork enters your mouth. A Web-based dashboard tracks your fork-based habits over time. You can also get coaching from the HAPI app.

The HAPIfork is not yet perfect, but it is an interesting step in the right direction. What's more, the company made the jump from research project at MIT to funded company at light speed. Now, with the product on the market, the company will receive feedback from real users rather than fellow researchers, which will enable it to continuously improve the product. Consumers must be included in the dialectic that surrounds the creation of enchanted objects.

Salt sentinel. As part of a health hackathon at MIT, I invented an enchanted saltshaker. Yes, it dispenses salt, if you really want it, and it is augmented with three capabilities: sight through a 360-degree camera, communication through a wireless connection, and data display with a pico projector. The idea is a riff and improvement on the food phone, which delivers the analysis of your meal by email or text message, an hour or so later. Too late to be helpful, unless you have a great deal of self-control or have prepared the meal well in advance. The saltshaker's camera does the evaluation in real time and the pico projector displays the analysis of your food directly onto the food itself, or on the tablecloth, so you can eat smarter as you finish your meal. There is also a strong social-feedback loop—you're encouraged to discuss the information with your companions at the table.

Bluetooth concept. This is design fiction, but given the trends in sensor miniaturization and wearables, it's feasible in the next five years. Imagine an actual tooth replacement that responds to chewing actions and is able to sense texture, temperature, and chemical content of food and drink. With your smartphone, just program it to help manage your intake in whatever way you desire—reducing sugar or salt, increasing vitamins or minerals, adding fiber, or cutting back on carbs. Bluetooth provides feedback, as well as alerts and encouragements. Bluetooth sends a text message to your doctor or trainer if you're doing well or poorly. It vibrates to slow down your consumption, emits an embarrassing sound to keep you from eating the wrong thing, or produces a noxious smell that inhibits your social interactions. In dire circumstances, it magnetically clamps onto a tooth above so that you cannot continue to eat. Yes, this last option sounds Draconian, but the current, popular method is far worse. Stomach stapling or intestinal reduction surgery is invasive, you run the risk of infection, and you will probably suffer reflux the rest of your life. You could even die from complications of diabetes.

WHOSE DATA IS IT?

New tools for measuring our bodies, their performance, and our behavior are breaking down the walls that separate our home from the doctor's office. This part of the quantified-self movement is often referred to as connected health. Connected health devices include weight scales that can sense edema (swelling), which can mean you are at risk of having a stroke, and are paid for by insurers; blood-pressure cuffs that report trends to a nurse; pulse oximetry clips to sense oxygenation of the blood for chronic obstructive pulmonary disease (COPD); spirometers for testing lung function and anticipating asthma attacks; blood-sugar meters for managing diabetes; ovulation-cycle tests; and many others.

A critical issue for the quantified-self movement is who gets access to all the data we now collect about ourselves and our daily lives: the food we consume, the number of steps we take, the medications we ingest, details of our blood pressure and weight, the minutes we feel great stress, hours spent each day working in Microsoft Word or flinging angry birds, the number of calls made to customers or text messages dispatched, hours of REM sleep, and a thousand other categories of information that reveal our productivity, passions, and distractions. Should we alone see the patterns? If others, who? How much of the data? Over what time period? Should they have the ability to recombine the data and compare it to that of others? To share it with others? Can we require that data be destroyed at a certain point?

Who pays for these smart devices and enhanced services? Usually it's an insurance company or your employer, and because they are paying, they typically assert their right to analyze the data, usually before you do, and sometimes exclusively. This is an asymmetrical power dynamic that I abhor. Your data should be yours and yours first. You should be given tools to understand the accuracy of the information and analyze what it means. If you choose to share it with your physicians or insurer to help them take better care of you, or prioritize their services, that's your prerogative.

Still, I have long advocated the power of enlisting nonprofessional caregivers in sharing data. A family member or close friend can celebrate your successes or encourage you when you need a boost. Competition spurs results. And doctors can take better care of patients they understand more fully; the idea that a doctor can see how you're doing is an important motivator.

A couple of years ago, my mother-in-law's doctor wanted to check for a potential arrhythmia. She was asked to wear a Holter monitor, which continuously recorded her heartbeat for three days. Interestingly, it had a dramatic effect on my mother-in-law's behavior. She knew her doctor would be reviewing the data, so she was careful to do what she thought the doctor would want and to avoid things that might affect her heartbeat. She ate more legumes than usual, avoided any R-rated films, and did not swear once in three days. Why? Although she was told that the monitor was recording only her heartbeat, the device made her feel as if her doctor were keeping track of everything she did: monitoring her diet, entertainment choices, and language. This influence, which psychologists call the sentinel effect, is a critical component of connected health—very like the famous Panopticon prison design that keeps hundreds of inmates under control even when not a single guard is in the watchtower.

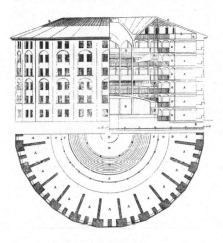

The Panopticon was a design for a hyperefficient prison where every cell is surveyed by a central guard tower. Today, doctors have the tools to survey patient behavioral data. Will this sentinel effect cause behavior change?

The Panopticon was the brainchild of Jeremy Bentham, the English philosopher (1748–1832). Its design allowed a small number of overseers, housed in a central core of the building, to observe a large number of building inhabitants. Bentham saw it as ideal for institutions such as prisons, hospitals, or schools. The inhabitants know they are being observed, but cannot see the observers, so their power and effect are greatly amplified—rather like the Oz effect. The durability of this effect declines as people acclimate to the building.

QUANTIFIED COMPARISONS

At Ambient Devices we experimented with a different approach to monitoring people's actions so as to encourage a positive change in behavior: being highly transparent by exposing data to a community and studying its effect on motivation. We conducted two experiments, both incredibly effective, one to motivate sales teams, the other to motivate exercise.

For a Bloomberg facility in New York we designed a set of ambient devices that would reflect the progress of financial transactions being made during the day. The device, a wheel eight inches in diameter, could spin on a base. The surface was printed with a spoke pattern that, when the wheel spun, appeared to spiral outward or spiral inward, depending on the direction of the spin. We mapped the direction and speed of these spinners to the real-time deals that five sales teams were responsible for executing throughout the day. They showed the relative measure of the daily performance of each team with a color-coded background.

We installed the spinners above the sales pit, so that about 120 people could see them and know, in an instant, which teams were ahead and which were lagging. You could not avoid seeing the data, but because it was encoded, you couldn't translate the speed to an actual number. This was important because visitors, often clients, came through the office and would not be pleased to discover their data was available for all to see. Because the spinners were installed side by side, we found that the salespeople—who are notoriously competitive—paid close attention to the relative speeds. The presence of the spinners had two effects. First, the members of a team were more likely to collaborate in order to beat the other teams. And they were motivated to work more creatively to make more sales.

Another experiment in motivating positive behavior came about as a result of my relationship with Richard Saul Wurman, the founder of the TED conferences, who was an early investor in Ambient Devices. I gave Richard an Ambient Orb for his swank Newport office, and he was so delighted with it that he said he kept "thinking of me" and would call me every few months just to talk.

In one of our conversations, I proposed a stunt that would show the motivational effect of side-by-side glanceable displays. At that time, Richard was launching a new conference called TEDMED to focus on health issues. He had put together a great lineup of presenters including Dr. Oz, Martha Stewart, Walt Mossberg (of the *Wall Street Journal*), and Naomi Judd, who had an amazing cancer-recovery story to tell.

A little team from Ambient showed up at the first TEDMED a few hours before the doors opened and installed seven color-shifting orbs on a long wooden shelf just outside the auditorium, where a thousand conference-goers would see them on their way in and out of the sessions over the three-day event. Below each orb we stuck a photo of one of the celebrities at the conference, and a color gradient that would shift from red to orange to yellow to green to blue according to how many steps that celebrity had walked that day.

When each of the celebrities arrived at the conference hall, I explained what we were doing and asked if I could insert a quarter-size sensor from FitLinxx (the precursor to Nike+) clipped to one of their shoes. Our receiver would connect with the sensor in their shoes and change the color of their orb to reflect how many steps that person walked that day. They all agreed.

What happened? Not only did everybody talk about these color-changing crystal balls, they started placing bets on who would "win" that day! Walt Mossberg and Naomi Judd took a stroll around Philadelphia instead of going to lunch, just so they could beat Dr. Oz and take the lead. The day's most important lesson was that public comparisons are incredibly powerful. Imagine a different approach. Would we have gotten the same effect if we had given our celebrities pedometers, without a side-by-side, pervasive display? I bet no one would have talked about the data or been motivated to change his or her behavior. Pervasive is persuasive.

THE QS MOVEMENT:
SELF-REFLECTION AND SHARING

The quantified-self (QS) movement is gaining momentum. Meet-ups in cities worldwide are attended by smart, curious, slightly obsessive experimenters who are measuring everything you can think of, from the age-old question "Where does my time go?" to more esoteric subjects including personal genome sequencing, psychological self-assessments, and automatic lifelogging—such as wearing a camera clip that takes a photo every thirty seconds and then compiles a time-lapse movie of your day.

I believe the quantified-self movement is in desperate need of more enchanted tools, including self-logging, spreadsheets, and scientific information visualization. These tools must include two critical elements: passive data capture and unavoidable display.

Today we have over twenty-five thousand health apps that, without passive capture and unavoidable displays, fail to help people change. They ask people to log their pill taking, for example, or their mood. These tools, even if they are well designed, aren't sustainable for people who lead busy lives. Few are still being used a few months after download.

But many enchanted objects are perfect, dedicated quantified-self tools that permit this important characteristic of passive capture. Nike+, for example, has an accelerometer embedded in the exercise shoe. Fitbit's activity sensor clips to a belt and sends data to the cloud so you can see your activity trends and compare them with those of others. A dedicated bracelet such as the Nike FuelBand, Jawbone UP, and Fitbit band also tracks activity. It's easier to remember to wear a bracelet than a pedometer clip. The FuelBand has a bright display on the band to show your progress and reward you with a little animation when you exceed your goal.

The downside of these wearables is they require recharging every few days, or they have no display on the device itself. We need new display materials like E Ink, and new parasitic charging strategies, such as motion and solar, to make these quantified-self tools into truly enchanted objects.

THE GLOWCAP: A WISH
FOR MY GRANDFATHER

Health and longevity are personal issues for me. My grandfather died too young of heart disease, and my father has struggled with it for years. This family history led me to think about how enchanted objects could help us with one of our most fundamental societal issues: health care, and in particular gracefully aging "in place."

This was the goal of my wireless health company, Vitality. After seeing the impact on awareness and behavior change from Ambient Devices, I was keen to bring these insights to a space where simple sensors, wireless technology, and glanceable data could make a big impact on people's health. I decided to focus on a single, small, but important aspect of health care: medication adherence. We already have amazing cures and medicines with proven efficacy on dozens of diseases, but every year millions of Americans fail to take their prescriptions. Almost half of the population of the United States is prescribed some medication by a doctor, but on any given day 50 percent do not take their pills as prescribed, often with serious health consequences and at an onerous cost to society—the New England Healthcare Institute estimates that this nonadherence results in as much as $300 billion annually in unnecessary costs in the United States alone.

The health-care industry has typically pinned the blame for non-adherence on patients. *If those patients would just remember to take our fantastic medicines, we wouldn't have any of these problems.* That statement has some truth, but the reality is more complex. If patients don't experience any symptoms, they may figure the medicines have done their work and simply neglect the full course. Or patients find themselves unable to get to a pharmacy for refills. Or they have trouble affording the prescription. Or they don't believe the medication actually works. Or they suffer unpleasant side effects.

In the summer of 2008, my business partner Josh Wachman and I asked ourselves, could an enchanted object make a difference here? If it could, wouldn't that have an enormous effect on our health-care system as a whole? We thought maybe we could change a daily behavior so that people would at least take their medicine consistently. We started out with a day-of-the-week pill container, added a glowy LED and sensor on the lid of each compartment, and connected the container to the Internet so it could send data directly to the pharmacy and order a refill before it was needed.

But after talking with people at the big pharmacy chains CVS and Walgreens, we realized that the day-of-the-week container was not going to work. We would have to use the standard pill packaging—the amber vial—because stocking and handling a nonstandard package would be too onerous. An enormously complex supply chain moves some 3.5 billion prescriptions annually in the United States. We could not expect pharmaceutical companies, package makers, freight handlers, and pharmacies to adapt their structures, technologies, and procedures to accommodate a new pill container, no matter how well designed it might be. We would just have to find a way to reinvent the amber vial.

We found the solution in the bottle's cap, which we could rethink and reshape without disrupting the supply chain. GlowCap is a smart pill-bottle cap that fits the industry-standard amber vial and contains a wireless chip that connects to the Internet via the Vitality Hub, which plugs into a power outlet like a night-light and can connect up to ten GlowCaps to AT&T's cellular network. The system can phone or text you if you've forgotten to take a medication. The cap reminds you to take the pill with a glow and an escalating ringtone. It also automatically sends a refill signal to the pharmacy when you're close to running out. And the caps can have their dose time and frequency programmed through a website.

DESIGNING FOR SUBTLETY
signaling must escalate slowly from subtle to insistent

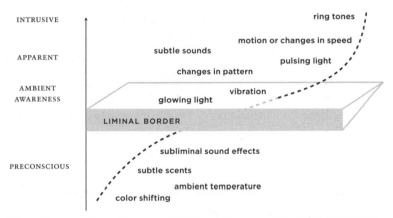

Think of the ways that objects communicate with you on a spectrum from subtle to more insistent. Ideally, enchanted objects shouldn't ever beep, buzz, or alarm. Instead they should respect your attention like a polite butler cleaning his throat to get your attention.

GlowCaps are a quintessential enchanted object: the form is an ordinary, everyday object, but it glows to remind you that it's time to take your meds, just as Neville Longbottom's Remembrall, in the Harry Potter books, glows red whenever he forgets something.

It's not just the cap or the glow or the connectedness that makes GlowCap work. As part of our research in developing the product, we were struck by how almost every successful program for developing new behaviors—such as smoking cessation, weight loss, and ending alcohol abuse—relies on social dynamics. The success of these programs comes from connecting the person with other people who care about them and can provide support in the effort to change.

So a "social nudge" became an important driving force behind the creation of GlowCap. In addition to the reminders, the cap also sends emails to friends or family so they can help support your medication adherence. It also transmits information to your doctor so he or she can keep track of how you're doing. The connectedness of the GlowCap enables a sort of Vitality buddy system. When my father and I are both using the system, I can see my father's weekly adherence history and he can see mine. This makes each of us accountable to the other, and more likely to do the right things for ourselves.

The reminder and refill features are important, but the social nudges are what make the difference. Involving a friend or family member is essential because it adds caring, motivation, and social incentives—you want to do well to gain approval and praise. The second nudge comes from your doctor, who becomes more engaged and tuned in to your day-to-day behavior. Especially for people who have a high regard for doctors and a healthy respect for authority, this connection can be incredibly important.

GlowCap is now a growing presence in the market and has succeeded in dramatically increasing the rate at which users take their medicine. People using GlowCap have an adherence rate of 94 percent, up from about 71 percent with a standard vial. The use of GlowCap is now widely considered to be a "best practice" in the health-care and quantified-self community—especially for patients with conditions such as an organ transplant, HIV, and diabetes, where medication regimen adherence is supercritical. The use of GlowCap also cuts the cost of care for insurance companies in a number of ways, by reducing the number of doctor and hospital visits and admissions, and avoiding the costly complications that can come with uncontrolled diabetes, such as limb amputations. GlowCap also lifts the revenue for pharmaceutical companies by as much as 30 percent or more, simply because more people take medications they are prescribed and refill more often. Everybody wins.

A LONG HEALTHY LIFE WILL SUFFICE

I would certainly be happy to live a good long time, but the reality comes down to that oldest saying—*as long as you've got your health*. As much as I rely on my trainer, trust my doctor, and look to my social network to help me feel well and strong, mentally alert, and physically capable, I also want a coterie of enchanted objects around me that are always available and tireless in their motivating effect so I can live as long and vital a life as possible. I only wish my grandfather had had the same opportunity.

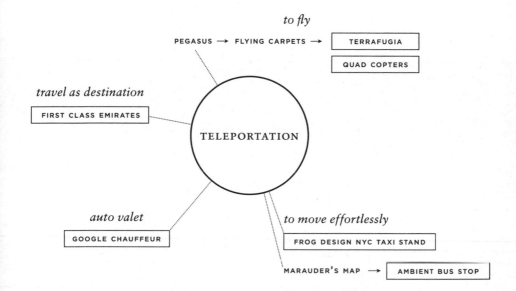

to fly

PEGASUS → FLYING CARPETS → | TERRAFUGIA |

| QUAD COPTERS |

travel as destination

| FIRST CLASS EMIRATES |

TELEPORTATION

auto valet

| GOOGLE CHAUFFEUR |

to move effortlessly

| FROG DESIGN NYC TAXI STAND |

MARAUDER'S MAP → | AMBIENT BUS STOP |

DRIVE #5.

TELEPORTATION: FRICTION-FREE TRAVEL

DAEDALUS'S DREAM WAS to fly. A character from Greek mythology, Daedalus ("clever worker") lived on the island of Crete with his son Icarus and was known for his abilities as an artisan and architect. He invented tools and constructed the great Labyrinth in Knossos, but his most daring experiment was motivated by his desire to fly. Daedalus studied the flight patterns of eagles, and then—so the myth goes—he fashioned a pair of wings from eagle feathers and wax.

When the moment came to test them, Daedalus donned one pair of wings and helped Icarus into a second set. Before taking off, the father warned his son not to fly too close to the sun because the heat would melt the wax and the wings would fail.

If you have traveled to Crete, you can imagine the joy of flying high above the islands surrounded by the azure Aegean Sea. Intoxicated by the view, Icarus could not resist soaring higher and higher until he came too close to the sun and the wax dripped and the wings drooped and then began to detach from his body. Icarus flapped harder and harder, but finally plunged to his death.

Fear of falling to earth has done little to dampen our desire to fly. We fantasize about and dream up all manner of mechanisms—from flying carpets to broomsticks, to flying cars and space elevators, and, best of all, teleportation. Teleportation means transporting matter from A to B without having to actually slog through the intervening miles of mud or traffic or outer space. One moment here, the next there. Once again, we have Gene Roddenberry and *Star Trek* to thank for the salient tele-portation-related cultural meme: *Beam me up, Scotty.* For all its advan-tages of speed and convenience, teleportation leaves out some of the pleasures of the travel experience—the wind in your hair, bugs in your teeth, surveying the landscape, meeting fellow travelers, encountering new sensations. If the journey trumps the destination, with teleporta-tion you've missed the good part.

But teleportation can sound pretty good in the modern era of air travel with all its travails. What should be an enchanting experience is too often anything but—a succession of frustrations, complications, delays, anxious moments, and getting lost. Today, most of us would be pleased to improve our travel experience—not so much in what hap-pens when we get wherever we're going, but in how we get there.

THE DIALECTIC: FROM FLYING CARPETS TO GOOGLE CHAUFFEUR

The dialectic sequence of teleportation devices has produced all man-ner of fictional and nonfictional devices over the ages. The Greeks were particularly fascinated by magical flight. Phaeton, the mythical son of the god Helios, had a truly enchanted travel facilitator—a flying char-iot—which did not have the operational problems of Daedalus's wax wings and upon which he races around the earth at his pleasure.[1] In the Middle Ages, Europeans transposed the chariot idea into a flying carpet. Exactly how a rug became imagined as a flying machine is a matter of conjecture. One theory is that Oriental carpets were expen-sive goods imported from exotic areas of the world,[2] such as Persia, where Europeans could imagine that magic reigned.[3] In the Renaissance, Leonardo da Vinci was known for studying how birds fly and for developing concepts for machines that could take flight, too.

In the nineteenth, twentieth, and twenty-first centuries, attempts to create flying vehicles abounded, alongside new and ever-more-outlandish fictional manifestations: gyroscope-powered monorails,[4] pneumatic "tube trains,"[5] cars with automatic navigation (proto-GPS?),[6] and hybrid "Aerocars."[7] The various iterations of airships included the zeppelin and the Goodyear blimp. In 1957, Hiller Helicopters designed a "flying fan" that looked "almost like a flying carpet."[8] The science fiction cartoon series *The Jetsons* (1962–63) picked up on the idea and presented us with a vision of space suburbia, with commuters zipping about in their bubble-topped spacemobiles. In 2005, Volkswagen introduced the Phaeton Lounge, a six-person limousine and homage to the mythical chariot.[9] Currently, the Terrafugia (roughly translated as "escape the earth") Transition is a flying car now available for preorder. In Danny Boyle's 2007 film *Sunshine*, a starship, *Icarus II*, is created to deliver a nuclear payload to the dying sun in order to save humanity.[10] The voyage is successful, but as in the myth of Icarus, it costs the voyagers their lives. The mythology of flying vehicles, while marked by failure, has nevertheless endured from ancient to contemporary times.[11] I often wonder what other fantasies and backyard attempts at flying have never been recorded and been lost in the mists of time.

Today, the effortless commute is getting some technological help from intelligent transportation systems, dynamically priced, load-balancing toll roads, and traffic-awareness devices, and on the horizon—and now a part of the urban zeitgeist—we see the driverless car.

I made a trip to the Googleplex in early 2013 to meet with Anthony Levandowski, product manager for Google's self-driving car technology, which Google expects to release within the next five years.[12] Critics of the program cite the complex legal issues, such as the prospect of a driverless car causing a fatal crash. But self-driving cars—combining ideas of the teleportation device, *The Jetsons'* car, and the first-class airline cabin—have the potential to revolutionize private transportation. And they are now inevitable.

With the fully autonomous car, road transportation will become more like riding the elevator or the train. All you will have to do is declare your destination, timing, and price sensitivity, and the enchanted device will take care of the rest. It will be waiting for you at the designated time and location, select the best route based on your desires (fastest, most scenic, or most efficient to pick up and drop off other passengers), and off you go—with your favorite tunes playing.

What will it be like to ride in a self-driving vehicle that does not need your attention to guide or manage it? You will be able to do all the things people already do while driving—texting, talking on the phone, eating, thinking, applying makeup—without the danger of driving into a bridge abutment. And you will be able to do many more things that you would like to do in your car now, but you are just too sensible, or scared, to try: writing, reading, skyping, making videos, holding meetings. When you're traveling on a good road without traffic, the sensation won't be one of teleportation, but it will be pretty enchanted what with all this newfound time. Before you know it, you will be alighting at your destination.

Because the car will be more of a service than an owned object, its relation to the individual will change. You will think of the car and the maker's brand differently from how you do now. You will not own the car or display it in your driveway, so you will be more free to match the model to your mood and your need. A sporty vehicle for a weekend outing, a vanlike thing for getting everyone to a family event. Splurge on a luxury car when you're feeling flush. Demonstrate your frugality by going for the cheapest model available. No longer will a single vehicle identify who you are or signal how much you're worth.

Then there is the safety issue. Today, in the United States, about thirty-five thousand people die in car accidents each year. Driverless cars will dramatically reduce this number because speed limits will be dynamic and lane changes and other road maneuvers will be controlled, based on traffic flow, weather, road conditions, and perhaps the price of energy futures. Self-driving cars will be far less prone to collisions for many reasons. They will never be drunk, texting, kissing, eating a burger, looking at themselves in the mirror, angry, late for a meeting, suicidal, excited by their new high-power propulsion system, or consumed by road rage. They will always be aware of the cars around them, the surface of the road, and will adhere to speed limits and road regulations. Accidents will be rare and the number of injuries and deaths will drop. We will spend less on emergency and health-care services, less on insurance policies and claims.

The technology of self-driving is already appearing and will continue to slowly spread into new models and products, just as other innovations leak into the market. We already see masses of sensors in high-end cars such as Volvo, whose rearview mirrors illuminate when they detect something in your blind spot. Mercedes and BMW offer auto lane detection and other features. Volvo has a new feature: it will drive around until it spots an empty space and parallel park itself. When you're ready to go, press a button, and it will come and pick you up.[13]

The 2014 Mercedes-Benz S-Class, which Mercedes has called "semi-autonomous," inches closer to the driverless car. The model is equipped with an onboard vision system that can keep the car driving within its highway lane and maintain a prescribed, safe distance from other vehicles at speeds up to 124 miles per hour. The driver can deploy the capabilities only in certain conditions, must keep his hands on the wheel, and must take over control when the environment changes, such as when exiting a freeway. In traffic, the car can track the environment and understand when to accelerate and when to brake.

THE ENCHANTED AUTO VALET

Another near-term goal in the evolution of the driverless car will be auto-valet self-parking. I don't mean that the car will execute the parking maneuver as you sit in the car, a capability that some cars already have. The auto valet will drive itself away after you've finished with it and proceed to the nearest or most optimal parking space, often much farther away than you would care to walk. This will reduce street congestion, and enable the placing of parking facilities outside of residential and commercial areas. While the auto-valet car is not being used, it will take care of cleaning, recharging, and fixing itself. Imagine all the time that will be returned to you when you don't have to think about fetching or parking a car or spend any time managing all the activities of car ownership and management—registration, inspection, maintenance, washing. Owners of today's electric vehicles have already had a taste of this delicious future, thanks to the elimination of oil changes and muffler replacements.

Skeptics of the self-driving and auto-valet car argue they will involve disruptions and modifications to numerous other technologies and infrastructures. As one blogger ranted, "If Google were really interested in saving people time and being beloved, why didn't Google focus on automating the DMV first?"[14] (Department of Motor Vehicles, that is.)

That brings up a critical point: enchanted objects are often only as enchanted as the systems within which they function. It would be irresponsible and foolhardy for me, or any company, to create a device that cannot carry out its task because its environment and social constraints won't allow it. We need to think holistically and systemically, as I'll discuss in the final chapter.

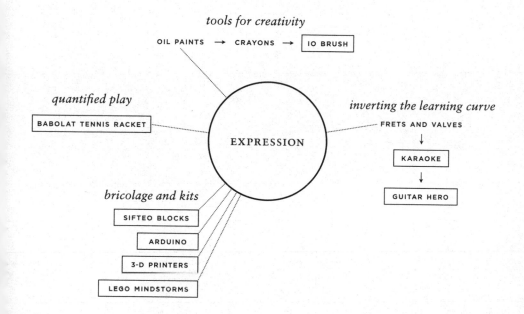

tools for creativity

OIL PAINTS → CRAYONS → IO BRUSH

quantified play

BABOLAT TENNIS RACKET

EXPRESSION

inverting the learning curve

FRETS AND VALVES

↓

KARAOKE

↓

GUITAR HERO

bricolage and kits

SIFTEO BLOCKS

ARDUINO

3-D PRINTERS

LEGO MINDSTORMS

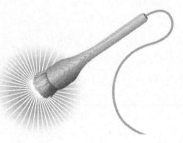

EXPRESSION: THE DESIRE TO CREATE

Do you know the old Chinese folktale about a kind young man, Ma Liang, who loves to draw and paint more than anything in life, but cannot afford paints? One night, the story goes, an old fellow appears to Ma Liang in a dream and bestows upon him an enchanted paintbrush. The next morning, Ma Liang awakes, takes up the brush, and immediately paints a fantastic butterfly, which, to his astonishment, springs to life and flutters away. This is no ordinary paintbrush! Ma Liang wonders if it might be put to better use than painting butterflies. He knows the farmers in the village are just as poor as he is, have little water for irrigation, and have a hard time nourishing their crops. If the paintbrush can bring a butterfly to life, might it work even greater wonders?

Ma Liang paints a stream flowing through the fields, and to provide a backup income source for the farmers, he sketches in a few cows. When Ma Liang's neighbors complain they are hungry, he paints some fresh fish for them. Soon enough, word gets out about Ma Liang's enchanted paintbrush. Along comes a wealthy landowner who steals the fantastic object in hopes of further enriching himself. He figures the simplest plan is to draw a mountain of gold, but the paintbrush refuses to follow his lead. The brush, it seems, has a strong moral compass. Figuring that Ma Liang has the necessary power, the landowner commands the young man to draw the golden mound. He does, but positions it on the far side of a great sea. The landowner, hugely excited, insists that Ma Liang draw a ship so the landowner can sail to the gold. Ma Liang does so and the landowner sets sail. When the landowner is well on his way, Ma Liang paints a great wave, which sends the landowner and his ship to the bottom of the ocean. Everybody lives happily ever after. Not only does our hero achieve his drive to manifest his creativity, he succeeds in providing safekeeping for his fellow villagers. Two fundamental human desires fulfilled with a few strokes of a brush.

The urge to be creative, to lose ourselves in the flow of being generative, is a primal human drive dating back to ancient civilization. We see signs of it in cave paintings in Spain, thought to be forty thousand years old, and in the extraordinary Lascaux cave paintings in France from over seventeen thousand years ago. Like Ma Liang's painting of a butterfly, these prehistoric paintings usually represent creatures of nature, particularly large wild animals such as bison, horses, or deer. We don't know why prehistoric artists chose to represent animals in their creative expressions—instead of other people or landscapes or whatever else was of interest—but some theorize these were works of magic, intended to give early men success in the hunt.

In addition to the visual arts, our desire to express ourselves often manifests itself in the creation of music. We have imagined many enchanted objects for generating melody and harmony—and the magical lyre of Orpheus is one of the most famous. Apollo, the Greek god of music, makes a gift of the lyre to Orpheus, and with it he produces music so beautiful and mesmerizing that it can charm wildlife, alter the course of rivers, and cajole trees and rocks into dance. On one occasion, Orpheus elicits music from his lyre so enthralling that it neutralizes the alluring, but deadly, music of the Sirens and saves the lives of Orpheus and his fellow travelers.

The lyre is best known for its role in helping Orpheus in his attempt to rescue his recently departed lover, Eurydice, from the underworld. Orpheus plays a song for the guardians of the land of the dead, Hades

The Ambient Orb is the simplest enchanted object. Essentially a one-pixel browser, it glows any color to reflect the status of the stock market, weather forecasts, traffic congestion, or just about any information you want to monitor.

My work on LEGO Mindstorms inspired me to find ways to embed computation and connectivity in everyday objects. The LEGO bricks contain touch, light, and proximity sensors, so kids can program their behavior through simple drag-and-drop programming language.

Sketches for shape-shifting, preattentive displays that we explored at Ambient Devices. If information can be expressed as a change in color, angle, height, puff, or wilt, then you can attend to the change you see in your peripheral vision. Glancing at an ambient display is much faster than reading text. It occurs in less than a second, with any cognitive load.

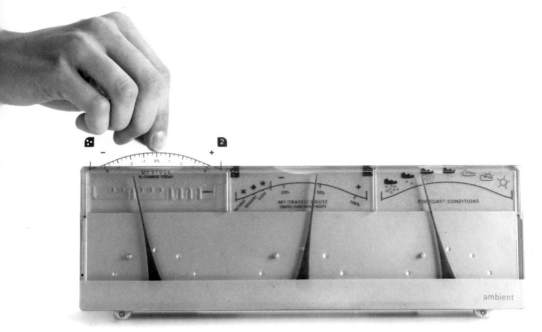

Swap in a card to physically program the Ambient Dashboard to show the data on the card: traffic congestion on your commute, stock market trends, or weather forecasts.

Cold 30s 40s 50s 60s 70s 80s Hot

The 5-Day Weather Forecaster automatically receives AccuWeather data on a dedicated display. Background colors shift to show you the current temperature outside so that you can read it from across the room without glasses.

Ambient Devices is now the leader in motivating energy conservation. The glowing color of the Energy Joule shows how much your home is spending on electricity. Multiple studies prove the presence of an ambient display effects a 30 percent reduction in home energy use.

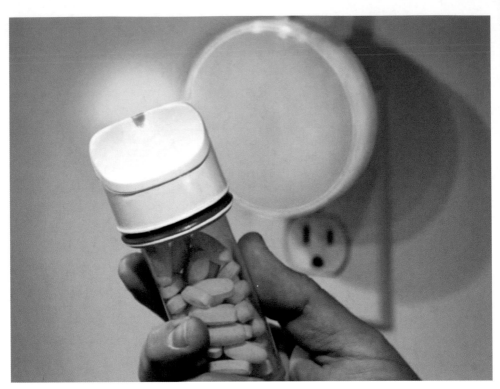

An ordinary object, augmented. GlowCaps
fit on standard childproof amber vials but
use AT&T's cellular network to provide text-
message reminders, weekly reports, and
automated refills to medicine takers. In a
Harvard trial, GlowCaps helped people take
their medication over 95 percent of the time.

Celebrating with Vitality cofounder Josh Wachman
at the 2011 Museum of Modern Art exhibit *Talk to
Me,* which featured Internet-connected objects,
including the GlowCap.

An enchanted salt shaker could provide feedback about what you're eating, right next to your food. Embedded cameras and pico projectors show information adjacent to the plate about what you are eating and where it came from.

This credit card prototype contains a battery, microprocessor, and accelerometer to provide financial incentives for getting enough exercise. It tangibly couples financial rewards with exercise. Earn for walking.

If your kids aren't willing to be tracked, maybe your dog will wear this Internet-connected collar.

The Goodnight Home helps remote family members feel connected. When a person in one location turns the large light on or off, the corresponding little light at the second location also changes state. It's a simple signal mapped to a natural gesture, and remarkably comforting.

Flower Power is a sensor that tries to keep your plants healthy by monitoring light, moisture, and the chemical contents of the soil. Houseplants that don't die? That would be magical.

The August lock is another example of old infrastructure that wants to be part of the cloud. Now you can unlock the door for a plumber or give your brother-in-law access for a few days, then take back their virtual keys.

Clothing, too, will become enchanted. Sensors in this sweater betray the excitement level of its wearer on a scale from calm blue to hot pink. Another dress by the same company becomes photoluminescent when tiny embedded cameras detect a spectator's gaze.

This enchanted wheel aims to encourage more commuting by bike. It senses how hard you are pushing, then redoubles your effort. A battery, electric motor, wireless controller, and regenerative brake are embedded inside.

and Persephone, which expresses his loss so powerfully it touches their hearts. They allow Eurydice to leave the underworld, on one condition—that Orpheus lead the way and not turn back to look at his lover until they both have reached the world of the living. As soon as Orpheus arrives in the sunlight he turns, but Eurydice has not quite emerged. At his one glance, she vanishes forever.

Enchanted objects animate love in full flower as well as love lost. In Mozart's opera *The Magic Flute*, the handsome prince Tamino falls in love with the beautiful Pamina just by gazing at a painting of her. "Dies Bildnis ist bezaubernd schön!" he exclaims ("This portrait is enchantingly lovely!"). Pamina's mother promises her daughter's hand in marriage, but there is a complication. Pamina is currently in the clutches of an evil sorcerer named Sarastro. If Tamino can free Pamina, he will be allowed to marry her. But how? The answer, again, is music. Specifically, a magic flute with the ability to transform sorrow into joy. Tamino's companion, Papageno, also equips himself with a set of magical bells designed to bring happiness to anyone who hears them.

As they make their journey, Tamino plays his flute and wild animals are so charmed they rear up on their hind legs and dance. In a plot twist, Tamino and Papageno are captured by an enemy. No matter. Papageno rings his enchanted bells, the captors cannot help but break out in gleeful dance, and the heroes flee. At last, Tamino rescues Pamina and returns home with her, playing his flute to protect them as they pass unscathed through trials of fire and water. (The Harry Potter tales pick up on enchanted music, too. Harry plays a flute, accompanied by his friend Quirrell on an enchanted harp, to lull the nasty three-headed dog Fluffy to sleep.)

YEARNING FOR MASTERY

We would all love to be able to bring images to life with our painting skills or charm others with music. From an early age, we are often convinced that we are doing just that. Think of the pleasure a six-year-old feels slopping watercolors onto paper or selecting the perfect crayon to sketch an elephant or a favorite monster. For most of us, the joy of unfettered creating fades away as we grow older and realize that creative expression—beyond a simple outpouring of emotion—requires technical mastery and talent.

So we often look to technology to enhance our skills and enable us to express ourselves without devoting our lives to the pursuit. In the visual arts, painting was made easier with the invention of watercolors, the crayon, and paint-by-number kits. In music, we found ways to make instruments easier to play. For stringed instruments, craftsmen added frets to the fingerboard to quantize and structure notes. To control the pitch in horns, we invented keys to control the flow of air and lengthen the pipes by a predetermined amount. In both cases, you trade expressiveness for simplicity. Baritone is easier to learn than trombone, but you have less fine pitch control. Guitar is easier to learn than the violin or cello, for the same reason.

In the past few decades, new technologies have enabled us to explore creative expression in media not imagined by the cave painters and flute players. In the late 1980s, for example, Bill Warner founded Avid Technology and invented the first digital, nonlinear editing system for film and video editors. Bill is a mentor and constant presence on the MIT entrepreneurial scene. Prior to Avid's breakthrough invention, film editors physically spliced and taped pieces of film together. Those who worked in video had to juggle multiple-tape cartridges and work with a machine that transferred the electronic images onto a master tape. Both processes were clumsy, time-consuming, and inhibited experimentation. Avid's technology revolutionized visual storytelling. It streamlined the editing of film and video by digitizing the process, permitting editors to quickly cut and paste elements together. From Hollywood to one-man wedding-video shops, Avid caught on like wildfire as old editing technologies—such as the Steenbeck, Moviola, and KEM flatbed editing machines—were pushed aside. Around the same time, a PhD student named Thomas Knoll sold a license to his new digital editing program, called Photoshop, to the Adobe Corporation. Two years later,

Adobe released Photoshop 1.0 to Mac users. In short order, Photoshop became the standard for manipulating digital images and the verb *photoshop* entered the cultural lexicon.

Avid and Photoshop gave artists and craftspeople invaluable, reasonably priced tools that permitted them to become more productive and creative. These technologies enabled artists to experiment with visual imagery in ways that hadn't before been possible. But you still need considerable training and practice to master these tools. Other enchanted objects, however, offer an experience more analogous to the child engrossed in crayons, LEGOs, clay, or the lyre of Orpheus or Ma Liang's magical paintbrush.

GUITAR HERO:
THE INVERTED LEARNING CURVE

Guitar Hero is the magic flute of our age. We all wanted (or may still want) to be rock stars. We play air guitar, sing in the shower, imagine competing on *American Idol*. The problem is, not everyone with rock-star ambitions has the musical talent to be the next Axl Rose. Learning a new instrument is intimidating and frustrating—for the first hundred hours, at least. My daughter has been practicing the cello for almost a year. Most nights, it's a struggle for all of us as we encourage her to practice, then suffer through the grating sound of mispitched notes and endlessly repeated phrases. Practice sessions often end in tears. It takes years before the learners can enjoy the fruits of their labor: playing with confidence and musicality.

Even if we're not prodigies, most of us want to fulfill that drive for creative expression, to be generative. So, wouldn't it be great to realize the fantasy of performing a favorite piece of music, in a crowded venue, and causing an audience to respond with pleasure and admiration? Or, more realistically, wouldn't it be satisfying to simply experience the joy of playing music without the intensive learning process?

By inverting the learning curve, *Guitar Hero* starts by giving you a feeling of mastery, then you work your way toward dexterity and musicianship.

That's the idea behind *Guitar Hero*. It's a game, originally released for PlayStation 2, that enables musical novices to experience proficiency and mastery—inverting the frustrating learning curve associated with playing a guitar, drums, keyboard, or other instrument.

That's largely because *Guitar Hero* features an enchanted object that looks like a real guitar but with some modifications. It has large colored buttons instead of frets along the neck. Instead of strings, there is a strummer pad. But the instrument looks and feels like a guitar, and even a first-time player instinctively understands how to hold the instrument and, soon enough, how to use it to make music.

Guitar Hero evolved from an earlier product, the *Axe*, created by a company called Harmonix, founded in 1995, by Alex Rigopulos and Eran Egozy of the MIT Media Lab. The *Axe,* released in 1998, allowed you to choose an instrument you wanted to play and control some elements of the music with a joystick. Tilt the stick up or down to select higher or lower notes, left for longer notes, right for faster eighth and sixteenth notes. The controller enabled you to improvise without making a mistake. You were a virtuoso, unable to screw up. The experience was amazing—wouldn't everyone want to play music if he or she had no fear of playing a wrong note or chord?

The joystick was fun and easy to use, but it had a serious shortcoming: it didn't capture the physical feeling of playing an instrument. You couldn't lean into the computer as you would with a guitar, rear back and shake your head with the beat, or leap up and down as you pounded out the final chords of your anthem. The algorithms were right, but jamming with a joystick just didn't capture the thrill of let-it-all-out creative expression.

What finally made the concept soar was the decision to couple the software with realistic hardware. The designers at Harmonix had already seen how successful this could be—they had developed a software-based game called *Karaoke Revolution*, which came with a real microphone. The computer played the music, you sang into the mike, and the software modified your pitch to make you sound as good as possible. Eran Egozy, CTO and VP of engineering at Harmonix, says that the beauty of *Karaoke Revolution* was that you instantly knew what you were supposed to do. Everybody knows how to hold and sing into a mike, even if he or she hasn't done it before. *Karaoke Revolution*, with its inclusion of a familiar but augmented object, paved the way for *Guitar Hero*.

During the development period for *Guitar Hero*, Harmonix put together a terrific team, including my company Interactive Factory and a group of skilled hardware designers from RedOctane. Harmonix perfected the on-screen, 3-D interface; RedOctane created the cool hardware instrument; and Interactive developed the graphics. Together, we selected the music styles, created characters, and added some narrative twists and turns to the software, including the possibility of getting booed off the stage.

We knew we had a great product but hit a few bumps along the road as we took it to market in 1995. A major issue was the box. Retailers want merchandise to fit neatly onto the appropriate shelf, and our box—containing the software and the instrument—was at least four times bigger than other software game boxes. One retailer after another said no. Finally, Best Buy took a risk on our brainchild and ordered eight thousand units. After just one day in the stores, Best Buy bumped up their order to eighty thousand units. The game struck a creative nerve, and soon millions of people wanted to experience it. According to the folks at Harmonix, who should know, it became the best-selling video game ever. It allowed people to realize their fantasy faster and also helped them to genuinely master the instrument—with enjoyment.

Guitar Hero does what an enchanted object is supposed to do. It leverages a familiar object, the guitar, so players can instantly interact with it in a natural way, even if they have never owned or played the instrument. *Guitar Hero* realizes one of our deepest fantasies—to be a rock star, a stage performer—and the software offers a support structure so you feel mastery quickly but can explore expressiveness and subtlety over time. The game is doing most of the work, but the crowd doesn't have to know that. They cheer you on when you do well and you feel as if you're in control. You deserve the standing ovation you get after finishing a song without missing many notes. It's emotionally exhilarating.

In addition to its success as a game, *Guitar Hero* had another surprising effect, which lovers of music should appreciate. You might assume the game would have lessened people's interest in learning the real instrument. Quite the opposite. After the game took off, the number of kids who started learning how to play a real guitar and other instruments increased. The core musical competencies of listening, rhythm, ensemble, practice, and confidence sped them on their way.

I'm surprised the *Guitar Hero* concept hasn't been extended to other instruments and forms of creative expression. Why is there no *Cello Star* or *Clarinet Virtuoso*? The concept of inverting the learning curve through providing more initial structure and playability could also be extended to other artistic pursuits—painting, furniture design, architecture. (This is what Instagram does with photos.) Often people envision artists starting from scratch with a blank canvas, but most artistic professionals have some form of guidance: clothing designers use patterns, screenwriters follow proven formulas, software coders use routines and libraries. Even Edvard Munch, the renowned Norwegian painter (famous for *The Scream*), created his paintings by applying paint over photographs. If the most celebrated of artists rely on enchanted objects for a bit of help, why shouldn't the rest of us?

MINDSTORMS: ENCHANTED TOYS

The many other forms of creative expression include the design and construction of machines, and this urge to engineer fantastic mechanistic objects was the inspiration behind LEGO's Mindstorms set of play products.

LEGO has been a sponsor of the Media Lab for many years. Seymour Papert, one of the founders of the Media Lab and a leading expert on the use of technology for learning, was involved in the development of Mindstorms—the name was taken from his book *Mindstorms: Children, Computers, and Powerful Ideas.* When the project was getting along in development, LEGO contacted me to see if the Interactive Factory could develop the software needed to accompany and support the toy. That led me into one of the most fascinating development projects of my career.

The people at LEGO had long been intrigued by the question "What if LEGO bricks were smart?" In the 1980s and 1990s, the question about how to add computing capability to almost everything was in the air at academic institutions, design firms, and product companies of all kinds.

The way LEGOs work, by snapping simple components together to make an incredible array of creations, is similar to how other forms of expression are created—including software programs and songs, which are built with simple, standard components. What if you could invert the atoms and the bits? Rather than snapping together software as if it were hardware, what if you could embed sensors and computation and code in something as small, adaptable, and friendly as the familiar LEGO brick?

That compelling idea changed my life in 1991 when I was a student at the MIT Media Lab—it changed my assumption that computers were destined to live forever as desktop machines. I realized that computation could be distributed, flexible, reconfigurable, and literally snap together. It was a profound first glimpse of ubiquitous computing.

Not only was MIT's work with LEGO a revolution in the way we think about computing, it was part of an even grander experiment in rethinking the nature of learning. The work of the MIT group was based on a constructivist approach to learning, and its members were deeply enamored of these ideas. Constructivism was pioneered by developmental psychologist Jean Piaget, who argued that knowledge is not a quantity to be transmitted from the mind of the teacher to the mind of the student. Rather, as Mitch Resnick, LEGO Papert Professor of Learning Research and head of the Lifelong Kindergarten group at the MIT Media Lab, puts it, "Learning is an active process in which people construct new understandings of the world around them through active exploration, experimentation, discussion, and reflection. In short: people don't get ideas; they make them."[1]

For the members of the Mindstorms development team, the coolest software programs were those, such as *SimCity*, that encouraged learning. For them, drill-and-kill games were old-school and uncool. The essential element of constructivism is the engaging project. I knew the benefit of project-based learning from childhood. As a ten-year-old Boy Scout, I entered the Mini Soap Box Derby. I remember developing theories about how a car behaves and debating my ideas with my father and my friends. *A heavier car travels farther than a lighter one. Small wheels accelerate faster. The rear end should be tapered to disperse the wind and avoid eddies and currents like in a river. Graphite power is the key.*

Today, many five- to ten-year-old kids I know are obsessed with *Minecraft*, a multiplayer, computer-based activity that involves "terraforming"—shaping and placing blocks to create all manner of things: houses, apartment buildings, whole cities. Kids are obsessed and wonderfully creative with it.

LEGO Mindstorms is a member of this toys-as-expression dialectic. During the development of the product, the belief in providing building blocks for kids to think critically and systematically about problems— even about problem-solving itself—proved to be a perfect fit between the MIT constructivists and the LEGO designers and leaders.

The breakout idea for Mindstorms, which emerged from the group mind of the development team, was to create a programmable LEGO brick that would also contain sensors that could respond to light, sound, and touch—so that the machines and creatures created with the bricks could engage with the real world in various ways determined by the toy's creator. The behavior of the system could be programmed with a drag-and-drop interface. (One version of the language, interestingly enough, is called Enchanting.) For example, if you wanted to make a bashful or timid creature, you could program a motor so it would immediately go into reverse when it hears a loud sound—like a crab retreating under a rock. The types and variations of the behaviors you could create were endless. The resulting creatures seemed incredibly alive in the way they could interact with other objects and with the environment.

By 1995, LEGO's CEO, Kirk Kristiansen, and his colleague, Torben Sorensen, who was vice president of LEGO Dacta (the company's educational division), thought the programmable brick was ready to be commercialized. That's when my company, Interactive Factory, swung into action, creating the video and CD-ROM of software examples that shipped with the bricks.

LEGO Mindstorms was launched in toy stores around the world in 1998. In the United States, it quickly became one of the hottest-selling Christmas gifts that year, selling eight thousand units in less than three months at $199, which seemed like an astronomically high price for a toy. The *New York Times* heralded Mindstorms as a "transition . . . [that] illustrates the evolution of technological play," bolstering a toy company that had been "overshadowed by video games and other forms of electronic play"[2] with a whole new product line.

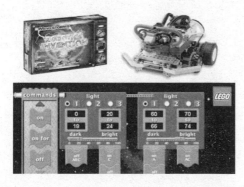

LEGO Logo is the drag-and-drop software to program your Mindstorms invention. It's a behavioral construction kit.

That Mindstorms kits appealed to the adult hacker community as much as they did to kids was a huge surprise to the LEGO leadership, who were nonetheless pleased to see sales climb by 300 percent in 1999.

My experience with Mindstorms furthered my thinking about enchanted objects. LEGO toys have always given people the ability to transform whatever they can imagine into a physical reality. Mindstorms took that idea further. It gave people the ability to program not just things, but things with any *behavior* they could imagine. A robot that knows not to fall off a table. A creature that, when it senses morning sunlight, proceeds to the nearest window and opens the shades. I came to think of Mindstorms as a behavioral construction kit.

Learning to code will likely become an essential skill in our society, just as reading and writing are today. Tools such as the Mindstorms Robotics Invention System make coding accessible and appealing to everyone. I'm not talking just about kids who have been brought up in the computer age. Even my mother, who is seventy-six, loves working together with her grandkids to put together new Mindstorms creations.

The drive for creative expression never gets old.

PART III

THE DESIGN OF ENCHANTMENT

THE EXTRAORDINARY
CAPABILITY OF HUMAN SENSES

WE HUMAN BEINGS have tremendous capability to perceive the world around us, thanks to the combined input of our five senses. It's curious then that we engage with technology primarily through just one of those senses, vision—and we scarcely use the entirety of our visual capability, at that. The other four human senses—touch, hearing, taste, and smell—with all their amazing abilities to sense information about the world are relegated to the sidelines. Today, we spend most of our technology interaction time staring at little glass slabs, which are positioned right before our eyes and in the center of our focus. This must change. We need to better understand the workings of all five senses so we can involve them more fully. This is particularly true when it comes to the sensing of subtle and subliminal phenomena—sensing, that is, on the borders of our consciousness.

To help in this department, we can look to researchers in the field of perceptual psychology and address some basic questions: What can humans sense? At what resolution? In what combinations? What are the limits of sense? How quickly do people activate their senses and acclimate to sensory information? How does sensual acuity vary from person to person? It's a fascinating field of study, and many of its findings can be applied to creating useful and innovative enchanted objects.

In Part III I dig into the process of creation and explore some of the fascinating challenges involved—the understanding of the human senses, the deployment of technological sensors, and the development of the most pleasing interactions between human beings and augmented objects—and describe a five-step "ladder of enchantment," a framework for the design of objects that will define the next generation of the Internet of Things.

THE COCKTAIL PARTY EFFECT: CENTRAL VERSUS PERIPHERAL

Let's begin with a simple experiment. Put down this book or glass-slab reader. Stand up. Hold your arms straight out to your sides. Look straight ahead. Can you see your hands? Not quite. Move them forward slowly until you can see them. Just a few inches, right? Your peripheral vision typically encompasses a 160-degree arc. This wide span of vision is an extraordinary human capability that can be leveraged—although it almost never is—in the design of enchanted objects.

Another experiment has been dubbed the cocktail party effect. (It doesn't have to involve cocktails. Any kind of crowded human gathering will do.) Psychologists in the 1950s discovered that human beings have a remarkable ability to monitor all conversations taking place among the people in a room—as many as twenty or thirty of them. You have no doubt experienced this. You are in the midst of a conversation with an acquaintance about, let's say, your recent trip to Norway, when suddenly, clear as a bell, you hear a familiar word spoken by someone else halfway across the room. It might be the name of a person you know and care about, the title of a book you have just read, the company you work for, a favorite hobby, a sports team you love or despise. You have not consciously been paying attention to that distant conversation, but that key word leaps into your field of auditory perception. This is a form of peripheral hearing.

We have a well of untapped visual and audio capability. How much?

How can enchanted objects take advantage of our capability for seeing so much that is not in front of us? Let's concentrate on vision, because it is our primary information-receiving sense. Research has helped us understand that vision is not neatly divided into foveal (the central focus) and peripheral. Retinal trackers show that our eyes scan the scene before us in a rather remarkable way. The eyes are in constant motion, flitting from, and briefly focusing on, one element of the view after another. The brain's job is to assemble, in real time, a coherent whole from the fragments of images delivered to it by our eyes.

What's particularly striking about this process is that the brain zeroes in on faces. Unlike other animals, humans can discern incredibly subtle variations in faces and facial expressions. This ancient capability has helped us recognize kin at a distance, quickly distinguish friend from foe, and separate the truth-tellers from the dissemblers. We are remarkably adept at gathering information from tiny movements of the mouth, eyes, and other facial muscles, from eye color and shape; from lines and wrinkles and scars.

You know this to be true, but Herman Chernoff, professor emeritus of applied mathematics at MIT and of statistics at Harvard University, devised a clever experiment in 1973 that gave scientific proof of what we naturally understand.[1] Chernoff's goal was to map a data set to a set of facial expressions, to see if people could remember numbers better when translated into facial differences. He wanted to work with information that subjects in the experiment would not have any familiarity with, so he chose data about rocks from a mine in Colorado. This was a random choice; neither rocks nor mining had any relevance to the experiment. Chernoff selected dozens of different data points and then generated computer drawings of a face that would represent each of those points. As you have probably guessed, each drawing was slightly different—in the shape of the mouth, the opening of the eye, and so on. Chernoff found that people could read the variations in facial information just as consistently and with as much clarity as they could tell the difference in numbers. They could make associations between types of facial "data" and cluster them into groups.

The lesson of Chernoff's experiment is that our ability to read facial data can be deployed to do much more than sense emotions. But what if Chernoff had combined the mechanistic skill of facial-data recognition with our ability to recognize, interpret, and respond to emotion? Suppose, for example, if one of the data points pertained to the gold content of a rock and the facial expression that corresponded to it was a greedy grin. Or, if the rock was of no value for any known human or commercial purpose, the face was completely deadpan.

Humans have an equally exquisite ability to distinguish sound—different types, variations in volume and pitch, sequences and combinations. What if we could leverage that ability more completely in creating enchanted objects? Just as people could process large quantities of data in Chernoff's experiment through their ability to detect subtle differences in faces, we can do similar things with hearing—map information to a huge range of sounds and melodies. We already do this in a rather crude and superficial way with ringtones, assigning a snippet of melody or a type of chime to the phone number of callers we know. There is vast unrealized potential here.

TAPPING INTO IDLE SENSES FOR MULTIMODAL INTERACTION

When it comes to using your senses, how much can you multitask? If you are focused on a vision-intensive task, how available is the auditory part of your cognitive pool? Can you listen to a news program or a basketball game, then compose an email or sketch a drawing and monitor your kids at the same time? Do these abilities vary between people? Is multitasking a type of fitness that varies with age?

When I was a kid, my dad and I would drive into the Canadian wilderness every summer for a week of fishing. It was a long journey from our house in Madison, so we spent a lot of time listening to audiobooks and old radio shows on tapes from the library, including my favorites, *The Lone Ranger* and *The Shadow*. I didn't think about it then, but it was easy for both my dad and me to pay full attention to the words being spoken and the music being played because our hearing capability was not required for driving. Dad's vision and sense of touch were fully engaged, but his audio channel was almost completely available.

The unused, or underused, sense channel offers us another design opportunity for enchantment. Most information workers have plenty of spare audio capacity, because they spend their days dealing with information through reading and writing—composing and responding to email, tuning financial models, editing documents and presentations. What further information could they absorb via the audio channel, before it became a distraction or the focus of their attention?

Georgia Tech professor Bruce Walker developed SoundScapes to leverage the latent audio channel. The concept is similar to Chernoff's computer generation of faces, but Walker used a wide range of sounds taken from nature to convey specific types of information about the stock market.

SoundScapes starts with a general theme, such as a forest. There is a constant bed of sound, perhaps the thrum of a flowing river. As a specific data value fluctuates, let's say the price of gold, discrete sounds within the general theme change—the sound of birdcalls intensifies, for example. If the price continues to rise, the system adds another sound and also randomizes them to keep them dynamic and fresh. If the value of gold declines, the system removes one sound after another. If the price tanks, SoundScapes brings in the sound of thunder, then adds rain.[2]

Even this system only scratches the surface of our ability to distinguish differences and receive information through sound. The most obvious example of the genius of our hearing is our ability to apprehend tremendous subtlety in music, although it requires training. Though high school and college I took private voice lessons, sang in the St. Olaf Choir, Trinity Church Choir, and the Tanglewood Festival Chorus, the group that performs with the Boston Symphony Orchestra each summer at the Tanglewood Music Center in Lenox, Massachusetts. I've worked with some of world's greatest conductors, including Seiji Ozawa, Simon Rattle, and James Levine. Their hearing capabilities, both focal and peripheral, are astonishing. In rehearsal, Seiji will drop his hands to his side, causing a combined orchestra and chorus—as many as two hundred musicians—to come to a screeching halt in the midst of a passage in Beethoven's *Missa solemnis*. He will then peer at the oboist and say, "Your B-flat was a little sharp." Or, gazing ruefully at my tenor section, he notes, "I need you to better enunciate the *c* in the word *credo*!" Or to the violas: "In measure twelve, your entrance was late by a quarter beat." These are subtleties and nuances that the untrained ear could not catch and that even trained musicians will miss in the midst of such a storm of sound. The lesson is that music contains a huge amount of information and delivers it straight to the conscious mind and also at the edges of the subconscious.

Along with hearing, touch is also an underused sense in the creation of enchanted objects, and I am constantly encouraging students in the tangible media group at MIT to use touch-based, or *haptic*, technology in their designs. I mentioned the jacket that hugs you. We have also imagined a phone that gets heavier as your voice-mail messages stack up, a shoe that nudges your feet to walk in a desired direction, a door handle that heats up to signal that people are in conflict inside the room, and a wallet that gets harder to open as you approach a spending limit you have set for yourself.

The possibilities are endless. Given our skin's surface area, how much information could a full suit of clothing—fitted all over with haptic sensors—collect and convey? If we can learn to read braille with our fingers, how much information could we manage over our entire body just by touching and tapping?

THE SCENT OF YOUR BRAND

And finally we come to the most underused sense of all: smell. Smell is a powerful sense that can be emotionally evocative and provocative, but scent is not an easy quantity to master and manage.

I recently met a woman who works for Chanel, the fashion brand, at a conference where I was speaking. She talked about how almost every environment, from retail to hospitality to schools and hospitals, is now being designed with scent.

Like wine tasting and assessing the quality of stereo speakers, aroma has its own vocabulary: floral notes, animalic essences, vegetal undertones. In an experiment at MIT, researchers created an Internet-connected scent spritzer, similar to the kind used to freshen up the air in hotel lobbies, and connected it to a real-time stock-market data feed. When the market trended up, out came a hint of mint. When the market went down, a lemon scent filled the air. Unfortunately, our noses are not as capable of distinguishing variations in scent as our eyes are of recognizing differences in facial detail. The smart scent spritzer could not deliver nuanced information beyond the rather obvious minty accent or lemony drop. Magnitude could not be expressed: a big blast of mint smells just about the same as a whiff. What's more, the human nose quickly acclimates to a smell, as we have all experienced. To refresh an aroma so that people would continue to smell it required the introduction of a huge amount of scent. After a few days of the experiment, the room reeked of a lemon-mint combination that no other aroma could break through.

The designers and engineers at Mercedes-Benz, however, are leveraging the power of smell to complement a superior driving experience. Many car models already tap into the other senses. They have soft cowhide seats and leather-wrapped steering wheels (haptics), burled-wood panels (visual), and noise-dampening glass and surround sound (audio). Might the driving experience be richer, softer, and even sound better with the addition of fragrance? Mercedes thinks so. Its 2014 S-Class model offers an optional "Air Balance Package featuring filtration, ionization and fragrance diffusion." The system's developer, Sabine Engelhardt, a futurologist in social and technological research for the German automaker, worked in collaboration with the company's "olfactory team," whose members were already working to create the ideal automotive "room scent." Engelhardt brought in Marc vom Ende, a professional perfumer, and the team came up with a perfume atomizer, enclosed within the instrument display, that diffuses scent into the cabin through the ventilation system. You can choose from one of four aromas: Freeside Mood (citrusy), Sports Mood (flowery), Nightlife Mood (leather and cognac), and Downtown Mood (musky).[3]

THE COLLISION AND CONFLATION OF SENSES

We know that creativity is often spurred by the collision of diverse elements and talents. Similarly, creative opportunities for enchanted objects can be found at the intersection, combination, and multiplication of senses.

Consider a military example. I spoke with a pilot who has logged many hours at the controls of the Apache Attack helicopter. While in flight, he depends on far more than just his sight: "I'm using a HUD [heads-up display] for situational awareness, navigation, and selecting targets, [radio] comms comes through on the noise-canceling headphones, weapons systems are an array of buttons over to my left, and I'm flying with my hands and feet. In some choppers . . . a bump-seat . . . shows the risk of enemy fire by changing the bumpiness on that side of my ass."[4] This pilot is relying on his sight, hearing, and sense of touch to fly his helicopter, communicate with others, deploy weapons, and avoid danger.

In the future, we will take cues from systems like this one and tap more into our five senses to detect signals and signal responses. Like a dancer—or pilot—the information worker will sense and signal by engaging all parts of the body. The potential for expressiveness in our interfaces is enormous. Imagine if we engaged our tongue, which according to my eccentric friend Jaron Lanier, the author of *You Are Not a Gadget*, can independently control eight continuous parameters.

Figuring out the right senses to employ and how to bring them together in enchanted designs will require complex orchestration—we humans are constantly confusing how our senses combine to perceive the world. We know, for example, that one sense influences another, and we tend to fold the quality of the senses together. For example, at Interactive Factory we produced multimedia exhibits for the Boston Museum of Science, the FDR Museum, and the Chicago Museum of Science and Industry, among many others. My clients would often review a finished animation that was accompanied by an unfinished "scratch track"—a down-and-dirty draft of the spoken words and type of music that would make up the final sound track. The client would comment and then, a few days later, review the animation again. The visuals had not changed, but now the sound track was complete and fully polished. "Wow," the client would say. "What did you do to the animation? It looks fantastic. Even better than last time! The colors are much richer now. I love the tweaks you made." But we had made no tweaks at all to the visuals. The client had confused one sense modality with another, as everyone often does. Improvements in the audio gave the illusion that the visual had changed, too. This conflation became so predictable that I learned not to reply, "Thanks, but nothing has changed in the visual. You're just conflating your sensory responses." I would simply smile, thank my client for the compliment, and let the client sign off on the "improved" design.

I often think of this experience when I walk into hotels with obviously designed odors or retail environments with upbeat music. The decor of the hotel feels more vivid, and so do the colors of the retail items that I'm now more likely to purchase.

Tapping human senses as a means to enchantment has huge potential, and we've come to expect it from staged experiences such as theme parks and restaurants, but there are limits. I believe that parallelism is a new muscle, a new fitness, honed and refined by the millennial generation. Younger people respond better to multimodal enchanted designs. I see my students gaming while tweeting, watching TV, writing a paper, and texting three friends. Their parallel ways are particularly obvious when I'm teaching a class at MIT. Thirty students, thirty laptops open. When I touch on a topic, they check *Wikipedia* for details. They scan through videos for a better view of a project or google the backstory of some inventor I mention. New enchanted objects will need to accommodate this type of parallel play across multiple channels and modalities.

TECHNOLOGY SENSORS AND ENCHANTED BRICOLAGE

Now THAT WE have considered human senses, let's turn the coin over. What about technology's capacity to sense? How can device- and object-based sensors be a part of enchantment?

BRILLIANT SENSORS

What can computers sense? The short answer is, a lot more than we can. They can sense sound, light, touch, many kinds of movement, biometric data such as heart rate and fingerprints, liquid flow, barometric pressure, radiation, temperature, proximity, and location.[1]

As amazing as human sensing capabilities are, computers are better than we are at sensing certain things, such as weights and measures, because they have infinite patience and can be engineered to sense for much finer degrees of precision. Humans can only see a small part of the visual spectrum or hear a limited range of frequencies. My father seems to have a compass in his head, but many of us would be hopelessly lost without our GPS and clear signage.

What, then, can computers not sense that humans can? Today, computers are unreliable at reading and interpreting subtle facial expressions, body language, and emotions—although affective computing is a hot area of research and algorithms are improving.

Still, we can embed many simple sensors in everyday things to make them enchanted. The least expensive of these are switches (what the GlowCap has inside), light-sensing diodes (what allows the SunSprite to measure light), and accelerometers. The Nike FuelBand, the wristband that tracks exercise, contains an accelerometer that senses footfalls and the dwell time of the foot on pavement as you run. This allows the band to measure pace and, thus, calculate distance. The wristband shows total steps, calories burned, and a new Nike-invented unit called "fuelpoints." On the other side of the sensor-price spectrum is Google Chauffeur (the self-driving car's software), which operates largely thanks to a $100,000 LIDAR, or laser-enabled optical scanning system. This incredible sensor bounces a laser array off nearby cars, buildings, and people to create a high-resolution, 3D map at sixty times a second.

In every dimension, in every modality, technology can sense more and at higher precision, certainly with more patience, than humans can—with the exception, perhaps, of smell and taste. Technology can see better in the dark using infrared or sonar (as dolphins and bats do). Sensors/silicon can hear with more sensitivity across more frequencies (such as low-frequency earthquake sensors). And they can sense in ways that are impossible for humans, such as antennas that tune in to radio or cell phone signals, and at speeds that are almost impossible to comprehend, such as a single-gigabit fiber-optic cable sending encyclopedia-size chunks of data around the world in an instant.

Unlike humans, man-made systems do not excel at combining senses. It's hard to build systems that can learn and also figure out what not to learn. It's also difficult to build systems that filter out data noise and glean salience from information. Man-made systems are less flexible and adaptive, and more brittle, than ours are. Designers for enchantment will thus leverage robots for their innate abilities such as infinite memory, patience, and precision. They will graft these to the genius that makes us human: synthesis and leaps of creative insight.

Enchanted objects will pair the capabilities of human and machine to deliver more. More omniscience, more telepathy, more immortality, more safekeeping, more teleportation, and more expression.

INEVITABLE TRADE-OFFS

Rarely, if ever, can we incorporate all the abilities of enchantment into a single object. In designing any product that must be manufactured—unlike our fictional sun chariots and flying carpets—trade-offs are inevitable.

The Millennial Net company specializes in a type of machine-to-machine communication called mesh networking. Conceptually it's awesome. Each node in the network is also a repeater so that information spreads the way rumors spread—each person passes along information to the other people who are nearby, and eventually everybody gets the message. There is no preordained communication path. It's all ad hoc, and with no strict hierarchy of connection, the network is more likely to keep on working, even under stress. Unlike in a chain or hub-and-spoke topology, you have no single point of failure.

I met Millennial Net's founder and chief technology officer, Sokwoo Rhee, at an Internet-of-Things meet-up in Cambridge, where he provided example after example of how wireless sensor networks are being deployed in fields from health care, energy management, transportation, emergency response, agriculture, infrastructure, and buildings, to the military. He underscored the issue of system trade-offs. A wireless network in the perfect world would have six important characteristics—reliability, scalability, range, power, data rate, security—but the laws of physics and information science limit perfection. As a network designer you inevitably must trade performance in one area for performance in another.

We had to make some important trade-offs while developing the GlowCap. Because we thought it was essential that smart medication packaging have at least a one-year battery life, we had to trade brightness, pulse rate (the cap pulses with light only every second for a maximum of four hours per day), and range (the caps had to be within a hundred feet of the Vitality Hub, which connects to AT&T's cellular network) so the battery wouldn't expend too much energy in too short a time.

What the Millennial Net founder didn't mention were two constraints that drive so much of product design today: size and cost. The iPhone, for example, trades off battery life—it can hardly get through a day without a recharge—for aesthetic appeal, specifically a superthin body. Apple can charge a premium for beauty and thinness, but the user pays twice—in the cost of the product and in the reduced usability. If Apple had traded beauty for battery life and made the iPhone just five millimeters thicker, would it still have been able to establish a price point above Samsung and other smartphones on the market? Perhaps not.

SEVEN ABILITIES
of ENCHANTED OBJECT

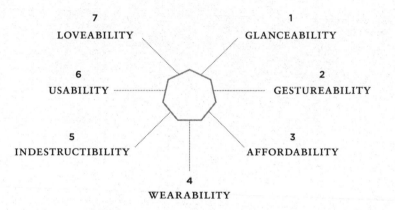

7
LOVEABILITY

1
GLANCEABILITY

6
USABILITY

2
GESTUREABILITY

5
INDESTRUCTIBILITY

3
AFFORDABILITY

4
WEARABILITY

THE SEVEN ABILITIES
OF ENCHANTMENT

ENCHANTMENT ARISES FROM a set of unique qualities that are sharply distinct from traits found in the other futures of terminals, robots, or prosthetics. I see seven "abilities" that differentiate enchanted objects from smartphones and their apps. This translates into how we learn them and they learn us. Their ability to engender trust, for them to act as respectful agents of our time and attention. The most important: glanceability, gestureability, affordability, wearability, indestructibility, usability, and loveability.

1. GLANCEABILITY

Enchanted objects help us make decisions, almost subconsciously, and bring information into focus at the most opportune time and place. When well designed, they lighten our cognitive load by giving us just the amount of information required to make the best choice without unneeded detail.

Glanceable from a block away, this bus pole concept helps riders pace themselves as they appproach the bus stop. It puts data at the decisional moment, is calming, and shortens people's perceptions of wait times.

THE AMBIENT ORB:
PREATTENTIVE PROCESSING

My inspiration in founding Ambient Devices in 2002 feels just as relevant today as it did then. The idea was based on glanceability, or what cognitive scientists call preattentive processing.

After college and before going to graduate school, I worked for Oberlin College for a year, writing software for professors. During that year I took a great course in cognitive science in which I learned about perception, attention pools, change blindness, the cocktail party phenomenon, how to hypnotize people, and deception in magic tricks. I learned that the brain processes certain kinds of information without having to pay full conscious attention to them—thus the term *preattentive*. Humans (and animals) subconsciously take in certain visual, auditory, and other cues in their sensing periphery.

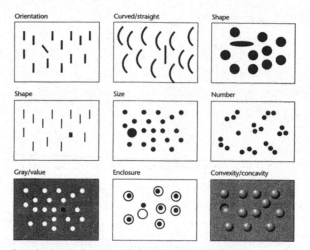

Certain visual phenomena are quickly perceptible by our reptilian brain. These preattentive examples are much faster to interpret than numbers of text.

This insight led me to invent the Ambient Orb, which I mentioned earlier, and found the company Ambient Devices with Ben Resner from MIT and Pritesh Gandhi, who still runs the company today. The orb is the simplest of smart devices, essentially an Internet-connected lightbulb. Packaged in a glass sphere, containing a wireless chip and microcontroller, and lit from within with diffused red, green, and blue LEDs—a mix designed to shift glow between any colors—the orb can track any kind of real-time data that you care about and that is available online. You might choose, for example, for the Orb to glow according to the pollen count that day in your zip code so you can be prepared with your allergy remedies, or to the wind speed to let you know if it is a good day for sailing, or to financial data, such as the movement of your portfolio. You can change the "channel" you track to match your needs or the season or whatever you care about most—such as this book's Amazon ranking. Simply go to AmbientDevices.com and select a channel.

The purpose of the orb is not to deliver information details. It has no number or text to display the precise pollen count, the wind speed in miles per hour, or the dollars you lost in the stock market today. Its job is convey a high-level reading in less than a quarter of a second. The orb is an enchanted object because it demands so little attention. Its single-pixel browser only conveys whether something needs your further attention. It answers the first question in the hierarchy of information processing:

1. Is the data worthy of attention now?
2. What is the information in summary? What's the headline?
3. What's the trend of the information? Is it getting better or worse?
4. Finally, what are the details of the data itself?

AMBIENT DISPLAYS RESPECT YOUR ATTENTION

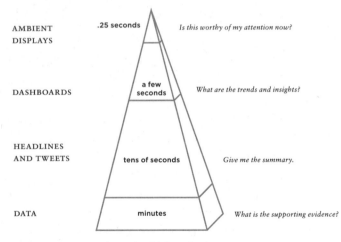

AMBIENT DISPLAYS — .25 seconds — *Is this worthy of my attention now?*

DASHBOARDS — a few seconds — *What are the trends and insights?*

HEADLINES AND TWEETS — tens of seconds — *Give me the summary.*

DATA — minutes — *What is the supporting evidence?*

time to absorb information

Ambient information displays don't require any effort or cognitive load. They can be perceived in less than a quarter of a second out of the corner of your eye. These displays help you prioritize your attention and are the only antidote to information overload.

If the answer to the first question is yes (worthy of my attention), then your brain transitions into a different thinking mode. Daniel Kahneman, winner of the Nobel Memorial Prize in Economics, writes about this phenomenon in his book *Thinking, Fast and Slow*. Kahneman defines two systems of thinking. System 1 is "fast" thinking: subconscious, immediate, involuntary, and essentially automatic, emotional in nature. System 1 thinking responds to certain kinds of information, such as colors, shapes, and smells, and is almost instantaneous—it takes about eleven milliseconds for our brains to register a bit of such information. That is so quick that you are virtually unaware that you are taking in information at all, and that is exactly how the orb works. If it is displaying the color that means "no attention required," you get the message without actually having to think about it.

When information is identified by the preattentive, System 1 thinking as worthy of further consideration, then Kahneman's System 2, or "slow" thinking, comes into play. This thinking is volitional, deliberate, conscious, more logical, and fully attentive. You are aware that you're thinking about the subject.

What we know from cognitive science is that certain kinds of phenomena can be perceived in parallel by your brain, and they are not mentally taxing to process, so you feel calm as you think, rather than jangled and intruded upon. Designs that are preattentive and can be "read" in an instant communicate through such characteristics as orientation, shape, size, clustering, enclosure, and color.

The Windows 8 interface, for example, uses these principles in the design of its tiles: it determines the most deserving information to display. It is more useful than the display of the iPhone, which is more of a navigation center and does not provide clues to any of the questions in the knowledge-processing pyramid.

At Ambient Devices, when we started sketching ideas for objects with glanceability, we tried out various features that could change orientation or size or shape. What if we had a plunger whose height would tell you about information? Maybe something that could puff out and pucker back in, like breath in your cheeks? Or maybe we could map three information streams at the same time, such as the key performance indicators of your business. What if the object could have "wiltiness" so that it seemed to droop when things were trending in a concerning direction?

At last we settled on the orb shape, a slightly flattened crystal ball, and color to summarize any one-dimensional dynamic information. People loved the object for its design elegance and functional simplicity, and also because its glanceability has, by nature, a calming effect. The orb is unobtrusive in your environment with the ability to instantly deliver information that is either reassuring or that gently suggests that you might want to attend to the matter further.

The orb does have a configuration interface, but it isn't on the device itself: users visit a website to determine what information to track, to identify the source for that information, and to determine the thresholds of information (at what point should the color shift?), what colors the orb should glow, and at what pace the change should occur (gradual shifting glow?). If we were beginning the design for the orb today, the interface could be on your mobile device and would be even simpler. You'd point your phone camera at the orb, then drag and drop the functionality you want onto it. Programming the physical world like this is known as augmented reality, or AR.

Enchanted objects are delightfully nondistracting because glanceability nestles into our preattentive, System 1, early-processing mode of thinking. Once we know the habits of the object (what information it provides, what color means what), which takes little time, the information becomes an easy part of our lives. When the orb glows amber, our stock portfolio is up, or when it glows purple, the pollen count is higher than usual. We absorb this information preattentively.

We are already surrounded by objects that, because they are glanceable, can provide summary information about deeper and more complicated information. A window, for example, affords us information about the time of day, the status of the weather, or, if it's an interior pane, the location of coworkers. A clock does this, too. Apps cannot. Apps need time to launch and navigate. An app almost always demands attentive processing, and the icon generally does not contain any status or summary information.

GLANCEABLE INFORMATION
LEADS TO BEHAVIOR CHANGE

In addition to its calming effect, the orb is simplifying. By separating the interface (which is on the Web, displayed on your screen) from the object itself, you are not being asked to handle information processing and device management at the same time.

The result is that people are not put off by the technology. They do not dread the interaction. They welcome the object into their lives and come to rely on it. They will look at the device instinctively, ten or twenty times a day, without any sense of frustration or annoyance. When that happens, if they can affect the variable, people start to change their behavior.

You have to be a little careful with this, however. I brought home an early version of the orb, set it up in the hall (just as my father had set up his weather station), and mapped it to our stock portfolio. My wife began watching it to the point of obsession. I'd come home and she'd say, "The Dow was way down today. We need to do something!" I'd call the broker and make some trades. It didn't take long to see that the orb was not always a benign influence.

What types of data are good fits for glanceable enchanted objects? An obvious one is energy usage, for which we have exquisite information, readily available, in tremendous quantity. As a society, we need to solve our energy-management problem, and part of the solution lies in getting people to change their consumption habits so we can smooth out the difference between the supply curve and the demand curve. If people changed their consumption behavior for the better, we could avoid brownouts, build fewer (if any) new generating facilities, avoid bringing new lines into congested places such as Manhattan.

As I discussed earlier, we adapted the orb for this very usage. We realized that an energy orb would be most effective if it could answer information-processing hierarchy questions one *and* two. (Should I pay attention? What is the data in summary?) To do that it would need to change color status and also display some actual digits. So we incorporated a data display into the ambient, glanceable object and called the product the Energy Joule, since a joule is a measurement for a unit of energy.

The first Energy Joule plugs into a kitchen outlet to provide real-time feedback on energy use in the home.

A second-generation Energy Joule is now being deployed by many energy companies. These companies give all their customers a Joule to encourage energy conservatism, especially during peak usage periods. We were able to reduce the manufacturing cost to less than $25, so energy companies can afford to give away the Energy Joule for free. The Joule is placed in a high-traffic spot in the home—usually the kitchen—glows the price of energy and shows how much is being consumed within the home at that moment. Without a glanceable display, this data would never be seen or acted upon. The Energy Joule has consistently shown that it can help people achieve a 20 percent reduction in their home energy use—these are industry-leading results!

2. GESTUREABILITY

Another inherent aspect of enchanted objects is that we instinctively, naturally know how to interact with them. They are familiar objects, augmented. They are graspable objects such as silverware, mugs, door handles, drawers, and cabinets. We sit in them, such as chairs and benches and couches. We walk on them, such as floors. These objects sense and may respond to our natural movements—a tap, a wave of the hand, or even, like Samantha's signature gesture in the enchanted 1960s sitcom *Bewitched*, a twitch of the nose.

We already know how to use tools, furniture, and appliances, so, with a dash of added magic, they can easily provide new features and services by responding to the gestures we already use.

Consider disposal, for example—throwing something away. That's a gesture that has been part of our lives for millennia. A toss, a flick, a gentle drop, into a wastebasket, a trash barrel, a compactor. But discarding usually has an associated need: replenishment of whatever you just finished. The Amazon Trash Can prototype, which I developed with Paul Franzosa, is an enchanted waste receptacle that brings the two needs together and is activated by that most basic of human gestures: the toss. The can is fitted with a tiny camera, so as you put an empty bottle or depleted box into it, it reads the object using computer vision or a bar code. The can is connected to Amazon.com, and the item is reordered automatically. If you don't want to place an order, toss the item into the can without going by the scanner. If you do place the order, then change your mind, give the can a gentle kick and it cancels the order. Another natural gesture. The trash can does not take care of all your shopping needs, but it is a nice step toward creating a home environment where your devices help you keep everything running smoothly—knowing what's available, what's missing, what needs to be replenished. Less list-making required, and, to use an industry term, fewer stock-outs.

The second version of this prototype climbed up one step on the ladder of enchantment by gaining an environment-friendly personality. As you throw out items, it proposes healthier alternatives. "Third box of cookies this week. Really?" "Coconut milk all the way from China?" "What about in-season strawberries?"

Image furniture to connect to remote loved ones. This cabinet is a dedicated portal to one other person. Just open for quick conversation.

3. AFFORDABILITY

Thanks to the falling cost of computing, new ways of thinking about embedding computing are now possible. In many cases, any incremental cost of augmenting a familiar object with technology is "lost in the noise"—that is, it is scarcely noticeable. Adding a camera and wireless sensor to the Amazon Trash Can, for example, costs less than a latte. The connected caps for GlowCap cost less than a single pill. Compare this to the latest iPhone, which sells for over $800 (without the onerous AT&T contract). Most enchanted objects don't need the latest quad-core chipset, vast memory to store videos, or retina displays. Enchanted furniture can be produced for little more than its dumb predecessors. The cost of the essential components needed for ubiquitous computing—chiefly connectivity—has fallen so sharply in recent years that companies can focus on enchantment without compromise. Cost constraints are largely gone. In some cases the cost of creating enchanted objects can be entirely subsidized because they induce valuable behaviors—energy conservation, the reduction of health-care costs, the purchase of more goods. The cost of an Amazon Kindle or Apple TV, for example, can easily be subsidized by the sale of books or movies or other content you are likely to buy over its lifetime. These companies should have enough data now to calculate an accurate customer-lifetime value for any new hardware, so I'm surprised more companies aren't giving their stuff away, especially to loyal customers with good credit scores.

The falling cost of hardware prompted me to develop a piece of cabinetry dedicated to a single person-to-person relationship. The prototype, which my kids use to connect with my parents, is called the Skype cabinet. It enables superfast, supereasy dedicated video communication between two people, using a kind of magic box that can be built into a cabinet or bookcase. All the components needed to make it function are inexpensive and readily available: a microphone and speakers of good quality, a diffused LED lighting source, a high-quality camera, a proximity sensor, a half-silvered glass mirror, a microprocessor that's fast enough for videoconferencing, and enough bandwidth to avoid lags.

I tucked one Skype system into a cabinet in our living room at the head height of my five-year-old. The camera sits behind half-silvered glass, which enables you to look directly into the eyes of whomever you're talking to, and the cabinet door becomes a diffused light-box. Using an approach similar to that of the LumiTouch picture frame, a proximity sensor illuminates the door when another person is near his or her device. With this device, your furniture is conveniently optimized for two-way visual connection with people you care about. Always there.

And always on. Because the cost of connectivity is so low, anyone can afford the Skype cabinet and leave it connected for long periods. Which of your friends, family members, colleagues, will also be interested in transforming a bookcase or kitchen cabinet into a moment for serendipitous videoconferencing? How many cabinets would you dedicate to special relationships? These decisions were easy in our family. My wife wants to talk more frequently to her three siblings. My ten-year-old daughter wants to talk to her best friend. I would want two, to chat more frequently with my sister and my parents.

The Skype cabinet is effortless to use. There is nothing to set up. It is just on. You quickly accept it into the functioning of your home and come to appreciate it not as a special connection but as a portal to the people you care most about. When your friend or family member is near his or her cabinet, the translucent doors on yours glow. You can open the doors and start a conversation, but you don't have to. If you do, the Skype cabinet makes the conversation as pleasing as possible. The sound quality is excellent and the lighting is flattering. You can pay full attention to the conversation or you can engage in parallel living—cooking or playing or moving around as you talk.

It is so much nicer than a screen-based video talk. You don't have to huddle around a computer. It's like opening your cabinet and finding a friend waiting inside. It certainly makes you think differently about furniture, and what we might do with all those bookcases after we stop piling up physical books.

4. WEARABILITY

Because technology is getting so small, and miniaturization is taking place so much faster than even futurists predicted, get used to the idea that we will be wearing technology—particularly sensors—in many more things we wear every day like shirts, buttons, and jewelry. Wearability is a way to liberate functionality from the tyranny of the black slab and distribute it into objects all around us and, indeed, *on* us. Distributed technologies are less complicated and more tolerant of human faults than terminal-centric ones because each function does less. For example, being less reliant on a single battery source means that if your earpiece fails, at least the mapping function in your glasses won't die, too.

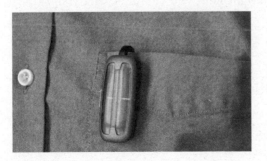

Like a pedometer for the sun, the SunSprite helps you get the optimal exposure to improve your energy, mood, and reset your circadian rhythm for a good night's sleep.

In 2012, I was introduced to two psychiatrists, Jacqueline Olds and Richard Schwartz, a husband and wife who teach at Harvard Medical School. They have a strong conviction that many people who suffer from seasonal affective disorder (SAD) or are prescribed antidepressants for other causes may be helped by bright-light therapy. It's hard for any of us to know if we are getting enough light. Light boxes can be helpful, but most of us prefer to get the rays we need from the sun. Jacquie and Richard conceived of a small, wearable sensor, not unlike a pedometer, that would measure how much sunlight you were getting during the day. As we started the design, we were always assessing the characteristics of the SunSprite, as we called it, as if it were jewelry. Should it clip to your clothing like an iPod mini or pin through like a brooch? Can it clip to a watchband? Would a man ever wear it as a pendant?

The design challenge was to make it very small, like a button, but also fit in the circuit board, sensor, and battery. Many small wearables suffer from short battery life. We've become accustomed to quartz watches that last years, but devices such as the Nike FuelBand need recharging every few days. The trade-off is size and data resolution. How often would the light sensor in SunSprite need to "wake up" and sense the brightness? Every second or so? Would it work well enough to sample the light once every thirty seconds or every minute? Decreasing this duty cycle could significantly prolong the battery life. Eventually we decided to use a piece of photo-voltaic material on the front of the SunSprite to give it indefinite battery life as long as you use it. By wearing it in bright light you charge it, and also recharge yourself.

NARRATIVE

One of the other most provocative examples of wearable technology that I've seen in recent years is Narrative, a "lifelogging camera." This small, orange, soft-shouldered object clips to your jacket or shirt. It contains a camera and is equipped with a GPS. It has no buttons or control bits at all. Without any prompting, it records a high-resolution "geo-tagged" image every thirty seconds—which means that it identifies each frame with your location, taken from the GPS sensor. Narrative is clever enough to figure out if you're lying down or standing on your head, so the app orients and organizes the images accordingly. With Narrative, you can track everything you did throughout the day.

The Narrative clip captures photos every thirty seconds to give you a time-lapse view of your day. How would your friends react if you started wearing it?

The metadata you gather with Narrative will be amazing. Imagine what you'll learn, what you'll remember. Who was that guy I met at the airport in Singapore? What was that delicious dish we shared some-time around September of 2013? How many minutes did I spend at the health club last year? How many hours writing this book or looking around Pinterest? Who was in that meeting when we decided to start the company? Who's my best friend based on total smiles per year? If you record long enough, you will end up with a visual record of (the rest of) your life.

Like an instant photo album, the Facebook Coffee Table shows your digital photos right when you want them most. Voice recognition understands keywords from your conversation and fades up relevant Facebook photos.

5. INDESTRUCTIBILITY

Enchanted objects can be almost laughably inexpensive and remarkably durable, much more so than an ¡Thing, which is optimized for portability. Like woodworking or metalsmithing tools, they can be made almost impervious to wear and last for decades.

I mentioned my days designing interactive exhibits for science and children's museums. The devices in those exhibits had to withstand the daily poundings and pullings, glee and abuse, of wave after wave of hyperactive schoolchildren, so I used industrial-strength hinges, military-grade buttons, and other hardened materials. I installed a major exhibit at the Chicago Museum of Science and Industry in 1995 and went back for a visit in 2012. The fancy Silicon Graphics computer workstations, which had cost $250,000 to purchase and had seemed so state-of-the-art at the time, now looked pretty unimpressive and even passé, thanks to the effects of Moore's law (the cost of processing power halves every eighteen months). But the devices, with their punchable buttons and unbreakable materials, were still going strong and even looked a little better for the years of use—with the patina and worn-in look of a well-used doorknob or much-handled tool.

Today's smartphones and computers, especially laptops, are far from indestructible. They are fragile, especially around kids with sticky fingers, juice boxes, and yogurt sticks. Fortunately, we can make interactive furniture such as the Facebook CoffeeTable much more spill-proof than iThings. This table listens to your conversation and recognizes the names of people and places tagged in your Facebook photos. When it makes a match, that photo is displayed on the huge, high-res screen under the coffee table. (You see, I am not categorically against screens.) It's like a visual track for your conversation. With a durable glass surface, bring on the spilled milk. I'm working on commercializing this smart table now for a major hotel chain as a self-service concierge service. The table features nearby shows and events, restaurant suggestions, local color and walking tours, and displays information about the status of traffic on the way to the airport.

6. USABILITY

The most delightful enchanted services work on your behalf, with minimal interface. Like an old chair, their use is self-evident. They have a patina of wear just like an old pair of gloves that proudly shows its heritage. They don't need an on/off switch and don't require care and feeding. An exemplar of usability is the Google Latitude Doorbell, which connects with your family member's smartphone and chimes when he or she is homeward-bound. To get the most value from it, all you have to do is remember which chime belongs to which person.

As we've discussed, our capacity for distinguishing and interpreting subtle sounds is remarkable. It's one of the five senses that we often ignore as interface designers, but the Google Latitude Doorbell makes perfect use of it. In the evening, after a long day of activity, while you're preparing dinner or getting ready to go out again, your visual channel is often saturated, but the audio channel still has plenty of capacity. That's why the doorbell is so simple and intuitive. The sound channel works in your cognitive periphery. You recognize who is on the way home without even trying. With the doorbell you don't have to text or call to find out when your family are getting home, you just know.

7. LOVEABILITY

Last but certainly most, enchanted objects must connect emotionally. Emotional engagement can be developed in a variety of ways—in formal cues, visual or gestural, that suggest anthropomorphism, or by using sounds such as the purr of a cat, or some of the communicative "gutturals" that small children emit to express glee or wonder or angst. These cues delight us, intensify our desire to bond with our devices, and invite us to use them.

Why is it hard to love a black glass slab? Perhaps its chameleonlike capabilities induce a type of category confusion—it is too many things to you. It's email and Chrome and games and shopping. Its multifunctional, multiuse, confused nature is too fragmented to bond with.

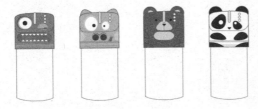

We worked with artists from Mimobot to make cute versions of interactive medication packaging. Take your medication regularly to keep them, and you, healthy.

Loveability results primarily by bestowing human attributes on inanimate devices, especially those with cute or infantile features. Konrad Lorenz, a Nobel laureate who studied animal behavior, argued persuasively that human beings react most positively to animals that have infantile features such as big eyes, big heads, short noses, and the like—neoteny rears its adorable head once again.

We can't make a device lovable by adding just any human characteristic. We do not want our objects to be boorish, loud, inconsiderate, intrusive, harsh, or rude. But they can be needy, such as Tamagotchi, who requires your ongoing care. The best way to charm people into a desired action—such as adhering to a routine (medicine compliance) or changing a behavior (energy usage)—is with a gentle touch or a soothing sound. Devices and services (and perhaps their creators) need to attend a finishing school for manners if we expect people to continue to happily live with them for extended periods.

As we become surrounded by more and more connected things (and we will be), then they must be aware of each other, wait for their turn to speak, and understand their relative importance to us. They must be unobtrusive, subtle, lovable, and polite for us to tolerate their collective presence in our lives—to make a more habitable near-future.

Now that we've done a "teardown" of the seven unique abilities of enchanted objects, let's build something.

FIVE STEPS
ON THE LADDER
OF ENCHANTMENT

FROM THEORY TO practice. What is the process of enchanting an object? What are the steps involved? Are there degrees of enchantment?

These are the questions and challenges that companies and designers must address in creating enchanted objects. On one hand, the explosion in the variety of sensors available for their use gives them more creative space to play in than ever before. On the other hand, they face rising customer expectations. People pay close attention to the look and feel of products. They care about the user experience. They are exquisitely aware of ease-of-use and drop tests, less forgiving of lackluster battery life. They know brands, relentlessly compare features, and will abandon the product with a poor review on Amazon or on a thousand early-adopter blogs.

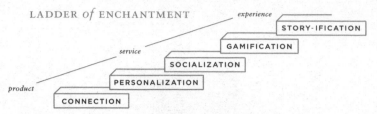

Climbing the ladder of enchantment bestows more personality, more product differentiation, and the ability to charge a premium with each step.

To help designers and their companies think through these challenges, I have developed a framework that I call the Ladder of Enchantment. This five-step, organized approach to thinking and creating products will deliver on the seven abilities of enchantment. This repeatable process demystifies the creation of enchanted objects or, at least, helps teams ask the right questions as they go. The higher the object climbs on the ladder, the more sophisticated or enchanting it becomes. Not every object need reach the top, but value can be added at each step:

1. *Connection:* adding sensing/sensor capabilities by connecting to the cloud.
2. *Personalization:* adding and leveraging personal information.
3. *Socialization:* adding connections to friends, loved ones, and colleagues.
4. *Gamification:* adding the fun and motivational elements of video games.
5. *Story-ification:* adding a human narrative for the product, service, or user.

STEP 1: CONNECTION

The Internet of Things is made possible through ubiquitous connectivity. An Internet connection allows the transmission of sensing and signaling information, the processing and storage of information, and the delivery of new services. We take for granted Internet service on high-end products—computers and smartphones—but we have only tiptoed into the possibilities for everyday devices.

Let's look at three pedestrian products—the bathroom scale, the kitchen trash can, and the office sign—to see how, in successive design steps, we can make them enchanting. What could be more mundane than these three things? Who would have thought to connect them to the internet? And why?

Let's start with the scale. When you step on it in the morning, what concerns are running through your mind? Your diet? Your exercise program? Clothing? Health? What effect can interpretation of weight data have on you? What services can a connected service offer? What magic might be possible?

Data on weight can change what you eat, shape your lifestyle choices, remake your wardrobe, spur treatments for health risks. If you're a diabetic or have heart disease, the scale could be a sensor with life-and-death implications. Can we turn a scale from an object of daily dismay to one of routine enchantment?

More simply put, how could a bathroom scale be augmented? Instead of merely showing you how many pounds you weigh in the morning, the scale could track and automatically send various measurements—such as weight, body mass index (BMI), body-fat percentage—to a document in the cloud, where they would be transformed into graphs and tables easily accessible to you, via the scale's mobile application, everywhere you go. The scale could even measure and monitor heart rate and stress level, and by connecting with your private medical history and the trove of medical databases online, it could predict life-threatening conditions. Of course, the scale would have to know you, and it would—by recognizing your unique pattern of biometric feedback or simply scanning your toeprint.

Now, the trash can. How could you infuse enchantment into the everyday, banal moment you open the kitchen trash can? What concerns cross your mind? Your environmental footprint? Your lengthening shopping list? Recycling questions? Packaging waste? What possible service could data collected on your waste stream offer? Your trash reflects your life choices, after all. The pattern of what your family eats and when. Your spending habits. Your disposal and recycling practices. Can we augment the trash can so it is transformed from a malodorous kitchen necessity into an item of enchantment?

Finally, the office sign. When you traipse around a large office building, trying to make your way through the maze of corridors and conference rooms, what crosses your mind? Finding the right person or meeting room? Getting A/V equipment or tech help? Locating an emergency exit or where to find food or a source of caffeine? Confirming meeting time and attendees? Getting word of room changes? Office signage directly impacts your daily productivity. What possible services could signs provide besides static labels and directions? What can be done to augment signs so they cease to be minimalist markers and become workday assistants?

I thought a lot about augmentation when I began developing the smart trash can that we have already discussed. What might a trash can do for its owners besides hold waste? Might it become a repurchasing system? What if the trash can had a scanner and camera in the lid that could read the item being thrown away and, through its connection to an online retailer, place an order for the same item?

What about the sign? I worked with the leadership team at Gensler, the world's second-largest architecture firm, with a staff of thirty-five hundred architects executing over three thousand projects a year, to create a spin-off signage business.

Signs are neither as boring nor as simple as they seem. Think of all the signs you encounter every day. Your college dorm, the hospital, airport terminals, office-building lobbies, hotels, elevators. Some buildings have thousands of signs. In many spaces, the use of rooms changes regularly, and facilities people need to change the information on the signs accordingly. The changes can be simple, such as what meetings are taking place in what meeting rooms, or more complex, such as when entire groups of people or functional units have been moved or reorganized within a building or campus of buildings. Signaling such changes can be a huge pain if signs need to be updated manually. And keep in mind that every sign in a public building in the United States must display, by law (Americans with Disabilities Act), information in Braille.

Gensler saw a huge opportunity for transforming the world of signage from a largely analog presence to a digital one. The idea was that smart signs could give people better guidance, instructions, and updates. They could turn deaf and mute signage into helpful tour guides.

Digital signage could make life much easier for building managers; they could simply update the display from an app. It would also provide a whole new landscape for the display of various kinds of additional information. In retail locations, signs could constantly be updated about merchandise, features, styles, inventory, pricing, and special offers. Signs could also be used to present commercial messages and advertising, where appropriate.

Connectivity, as a form of augmentation, does not necessarily deliver enchantment on its own, but—as with Gensler's signs—it can be the enabler of enchantment. To add connectivity, objects need a Wi-Fi link to the cloud through a router, a Bluetooth pairing with a phone, or an embedded cellular connection. Connectivity also requires the availability of reasonable pricing from phone service providers. We are now seeing this as telecom companies aggressively pursue relationships with device makers. Radio can be embedded directly in cars, coffeemakers, and dog collars—with data-plan deals of less than $1 per month.

Cellular connections hold the most promise because they eliminate the fiddling needed for Bluetooth pairing. They also allow users to avoid dealing with Wi-Fi networks, which, run in secure mode, require a password, a display to log on, or a USB cable or other tether, and setup time. The GlowCap bottle cap, shipped with an AT&T cellular connection, shows what's possible.

STEP 2: PERSONALIZATION

The next step up on the Ladder of Enchantment is to use the wealth of data generated by observing how large numbers of people behave to tailor unique services to individuals. Amazon offers you recommendations based on what other shoppers like you have purchased. Netflix does the same for movie recommendations. Google reads your Gmail and builds a profile of your interests, and those of your friends, to better target each of you with advertisements. Each of us has different preferences, motivational styles, and goals. When we connect an enchanted object to the cloud, it can start to treat us differently and tailor our relationship and its services.

Consider the bathroom scale again. If the scale knows our age, health profile, family history, and medications and understands our sense of humor, it can persuasively issue advice on the spot. It wouldn't bark out canned guidelines from, say, the American Heart Association. It would give us specific guidance based on our blood-cholesterol trends, history of strokes and heart attacks of relatives, the drugs we use for cutting LDL (bad) cholesterol, or diuretics we take to control heart disease.

Or think about the smart trash can. If the trash can knows our consumer preferences and purchase history, it might issue recommendations for fresh purchases of various consumables. Or it might issue alerts about new products just coming on the market that might fit into our diet or hobbies.

And let's come back to Gensler's smart signage for a moment. If the sign knows the visitor and meeting schedules for all conference rooms, it might send alerts to signs in the lobby to direct newcomers to the right room. Or it might post notices on rooms of upcoming agendas, meeting delays, or when participants will arrive. Meeting managers could tweet or text a message to a sign outside their offices—*back in five minutes*. Signs in hospitals could display data such as patient name, allergies, or risk of falling.

To create such signs, Gensler looked for new technologies, which is what brought them to the Media Lab and led to my involvement with the company. One of the major issues they needed to address was cost. Currently, digital signs are incredibly expensive, around $2,000 each. So digital signs can only be a success, from a commercial point of view, when the cost comes down enough so that they are seen as just another necessary building component, like doors or lighting fixtures.

To achieve this vision, it's necessary to think of pixels as paint—computer displays should be just as easy to work with, change, adapt, and apply as paint is. Preferably easier. One of the technologies that will make this possible is data projection, which we work with constantly at the Media Lab. The cost for a small data projector (known as a pico projector) is around $100 today and dropping fast, even as the projectors improve in brightness and resolution. At MIT, the Tangible Media Group created an I/O Bulb (I/O for input/output) that can project information on anything: paper, walls, tables, ceilings, and so on, creating what we refer to as digital "shadows." The possibilities of digital projection are endless. Objects could display directions for their use. A light over your dinner table could display nutrition information, next to your plate, of the meal you're about to eat.

Personalization could play an important role in our fragmented world of health care. Microsoft Health, for example, hopes to become a central repository of health informatics and then provide a service of analytics and diagnosis. The system would collect data from dozens of devices—glucometers, blood-pressure cuffs, heart-rate monitors, oximeters, thermometers, pedometers, stress watches, and others—that track your health and activity. Using the disparate data, and some powerful analytics, this service could keep an eye on your health progress and risks.

The service could also allow doctors, nurses, family members, or other caregivers to access the data and offer you help when you needed it. The advice would be highly personalized, delivering specific suggestions based on more data than you—or even your doctor—could be expected to know or remember. Suppose you return from Asia and, a week later, experience symptoms such as fever and aching joints. Your doctor in suburban Chicago diagnoses seasonal influenza, but your monitors analyze your symptoms in the light of a much larger quantity of data. Diagnosis: malaria. A very different treatment required.

The risk is that such a system could feel (and be) Big Brotherish. For individuals, subscription might be a requirement for employment in a big company, and as a result the caregiving could feel more as if it is being done *to* you instead of *for* you. The trick for the designers is to make sure these systems give people the sense of greater—not less—control and mastery over their health care. What is your weight goal? What is your glide path to that goal? How do the enchanted objects nudge you in the right direction to succeed?

For companies, especially large ones with tens of thousands of employees, these systems could help improve the health of their workforce and save millions, if not billions, of dollars a year on health care. They could help with pre- and postnatal care, reduce ER visits by identifying emergencies in the making, improve medical compliance, and more. To keep such systems in the realm of enchantment, rather than in the Big Brother zone, the information must remain under the control and ownership of the individual, not the company.

STEP 3: SOCIALIZATION

The third step on the Ladder of Enchantment is socialization—the connection of data with people. We already experience some sense of enchantment by sharing our stuff with people on Twitter, but we can create enchantment of a much more special kind by tailoring social messages to specific pools of people—loved ones, neighbors, colleagues, professional helpers of various kinds.

"Connecting with others" will come to have multiple meanings to designers of enchanted objects, because the "others" may well be objects. The connection may be between person and thing or between thing and thing, sometimes with a person as a go-between.

In the Thingiverse, objects can be treated as equal to humans. They can post, tweet, or swap data. They may connect with us using their own voices. In the garden, plants will call for water when they need it, inform us when their fruit is ready for picking, sound an alarm for the dogs to chase away marauding rabbits or hungry birds. Your recycling bin and solar panels will post their data and compare your progress with your environmentally minded friends.

Look again at the smart weight scale, and you can see the promise of socialization. Let's say you decide to allow the scale to deliver the information it collects to a network of fellow diet-conscious people who have joined a site and also agreed to share the information from their scales. The information is kept on a site accessible only by you and your group. You now have a tremendous new resource to draw upon in helping you achieve your goals. You can check in regularly to chart your progress, get feedback on your eating habits, and keep yourself accountable—while getting encouragement and support from others. If you start to flag in your efforts, you can get a "hard reset" from the members of your network, sufficient to make you recommit to the plan.

And what about the promise of socialization for the smart trash can? A scan of our refuse could trigger recommendations for future purchases—more healthy, green, socially responsible, cost-effective, and so on. Say you have committed to buying local ("slow") food. The trash can could issue useful messages about food provenance. *Did you know your coconut water is sourced from Malaysia?* Or, as your trash can learns about your eating habits, it could make suggestions based on market availability. *Maine blueberries are now in season.* You could share the data with others on social media to stimulate similarly responsible actions by your friends—and get support for your own. Nike+ does something similar. It allows friends to click and "cheer" during your workout. When they do, you hear a crowd roar with approval. We all want approbation for our efforts, and the socialization of enchanted objects can provide more of it, for more of our activities, than ever before.

STEP 4: GAMIFICATION

Once an object is connected, personalized, and socialized, the next step is to enchant users by getting them "in the game." The question: How do we get people off the bleachers of passivity and onto the field of action? How do we get them into the fantasy world of players and not spectators? With gamification. Gamification borrows from the tropes of video games to get people engaged even with mundane objects. It taps into our appetite for competition and progression, our dream to climb to the top of the heap, our wish to grow to and revel in some level of expertise and mastery. Who doesn't feel enchanted when circumstances and a little bit of effort sweep us into the limelight, if only in one small corner of the Internet?

Designers can add features to objects that video-game designers already use to motivate us: points, leaderboards, leveling up, streaks, and badges. Just as socialization taps our desire for nurturing and social comparison, gamification taps a range of drives such as achievement, recognition, and even a thirst for Vegas-like dopamine fixes from the anticipation of winning.

Points are among the most basic features of gaming. We all like to score and push our score higher, receive rewards for skill, effort, and persistence. The same is true for accumulating badges. With leveling up, we enjoy the recognition for more experience and the reward of fresh abilities at successively higher levels. Leaderboards then reward us with acclaim by peers, while we continue to vie with rivals for top spots. With streaks, as in gambling, we can be cajoled into capitalizing on—not cutting short—our progress and momentum.

How could gamification enchant the bathroom scale? We might receive points for weight loss (or gain) and long-term control. Or badges for accomplishing each milestone on our weight-control plan. On that opt-in social network, names might appear on leaderboards to highlight top performers in our weight-loss group. We might receive rewards for weight-loss streaks, urging us to keep with the program— and earn more rewards for persistence. "You've lost at least a pound a week for seven weeks," the scale says. "Why would you break this streak now?"

As for the smart trash can, designers could turn a family's buying and disposal habits into creative contests. Imagine consolidating family data on a social network to allow comparisons across entire neighborhoods or communities. Neighbors could team up with neighbors. Can your side of the street beat the other side? Can the Smiths or the Joneses buy more local, more green, with less packaging and more recycling? "Whoops," the trash can says, "that bottle is made of PET. You can recycle it—and earn enough points to get on the community leaderboard."

With signage, designers could play a key role in helping building managers increase room-use efficiency. People could be given points for arriving and starting on time, as well as for wrapping up as scheduled. Meanwhile, the room itself could issue signals to help occupants perform. If a conference-room reservation was expiring in ten minutes, but no other meetings were scheduled there, occupants would hear a ding ding—*it's okay to keep going*. If another group had booked the room, however, a different sound—ding dang—would indicate that it's time to pack up.

Today, many people are rushing to acquire sensors for measuring and quantifying their performance well beyond their weight and how they handle trash. The quantified-self movement,[1] which started with people tracking daily steps (using pedometers and Fitbit) and sleep (Lark, Zeo) and medication adherence (GlowCap), now involves much more: diet, learning, lifelogging, money, and mood state. Gamification features will accelerate this trend, as people vie for enchanted moments on the leaderboards or winner's circles of personal-habit sites. Socialization of the data will add to the excitement. Imagine the thrill of your children hitting the top rungs of Internet leaderboards for brushing their teeth.

STEP 5: STORY-IFICATION

The final step of the Ladder of Enchantment is creating or adding to a story that will enchant the user. Why a story? We all think of our lives as stories, each with a main character (us), theme, and plot (interesting so far, but as yet unfinished). We also love to hear stories about others and even about things. Stories hook into our curiosity—what happens next?—and into our emotions: What would I do in that situation? Stories have the unique power to engage and, if they engage enough, to trigger empathy, enchant.

Designers, having tapped the potential of personalizing, socializing, and gamifying, can work to embed a drama in our heads. They can involve us in the story so the narrative gains a purchase on both our minds and hearts. It becomes part of our heritage, our folklore, our mythology. We can feel as if we are part of the action, even a central character in the tale.

How would a designer story-ify the bathroom scale? Consider the possibilities in your own life. If you are on a diet, the goal of the diet suggests a personal quest. This may be small: you want to shrink enough to fit into a suit that used to fit perfectly or to get ready for the summer season in your swimwear. It may be more practical: to meet fitness standards for your job as a security officer. Or it may be much grander: to control cholesterol levels to meet a grandfather whose life was cut short by heart disease. By revealing your goals to your enchanted object, you allow the object to help tell your story—and help to make it come out with a happy ending.

We are always writing screenplays about and for ourselves, about who we are, what is happening in our lives now, and what we want to achieve. An enchanted object can add new scenes to the script, helping us to reaffirm and forward our quest. The bathroom scale, as mundane as it is, can play a role in our heroic journey to live a healthy life. The quest to maintain health, especially as we age, demands no less strength and courage than battles at other stages of life. An enchanted scale can play a role in the story line.

Product makers can take advantage of story-ification by imbuing objects themselves with stories. Even a smart trash can can have a life story. Remember Oscar the Grouch, the character on *Sesame Street*? When he made his first appearance on the PBS show in 1969, he gleefully sang, "Oh, I love trash!" Suddenly, Oscar's trash can was an object of great interest. Similarly, your smart trash can could become a member of the household, just as the barometer was in mine when I was a kid. Is the trash can a nag, constantly trying to keep you from buying more cookies? Is it a brilliant globalist that knows everything there is to know about food sourcing? Maybe the trash can is your partner in financial management, constantly working to save you money by suggesting different brands, alerting you to promotions and special offers, or advising you not to reorder an item that you regularly throw out without even opening. Just as we have come to talk to our computers and think of our cars as our friends, there is no reason that an enchanted trash can could not be endearing.

By climbing to the top rung of the Ladder of Enchantment, designers will come up with fresh ideas for engaging users with objects of all kinds. The route to enchantment will not come from brainstorming along just one avenue of enchantment. It will come from a synthesis: connecting, personalizing, socializing, gamifying, and creating stories. Deploying one or another, or perhaps two together, may lead to the creation of a cool product. When you bring them successfully together, people will be enchanted by the things around them as never before.

PART IV

ENCHANTED
SYSTEMS

Now, LET US more fully connect.

So far I have talked mostly about discrete objects: smart and connected, yes; members of the Internet of Things, for sure; but still discrete and largely functioning on their own. Tools and toys. Appliances and accessories. Vehicles and vacuum cleaners. A world filled with these enchanted objects will be a much more delightful and workable place to live than one covered by black glass slabs of all shapes and sizes.

Yet we can reach an even higher level, a broader vision of enchantment, and the best way to describe it is as *enchanted systems*. Not just trash cans and bathroom scales and doorbells, but an enchanted home. And more than benign surveillance, smart signage, and pens that remember meetings, imagine enchanted workplaces. Not only driverless cars and bus stop signals, but enchanted communities, towns, cities. Even whole societies.

You undoubtedly know something about systems thinking, even if you haven't studied the science or theory behind it. Two elements of systems thinking are important to the idea of enchantment: systems are self-regulating through feedback loops and are complex, with multiple components that interact with and affect one other, and that—to some degree—have the ability to learn and change and adapt. The human body is a self-regulating, complex system, as is nature as a whole. A rock, however, is not a system.

Gradually, systems thinking—derived from the study of natural ecosystems—has been applied to structures such as business organizations and social communities. And now, systems that do not include human components, such as cars, are getting closer and closer to becoming complex, self-regulating, adaptive systems of multiple components.

We know what an enchanted system might look like—it will exhibit some or all the characteristics of enchanted objects we have so far discussed: ubiquitous and constant connectivity; a focus on fulfilling fundamental human drives; natural interactions between humans and their tools and services; technology receding into the background fabric of life.

Enchanted systems will work on a grander scale, however, than just a connected Internet of Things. In an enchanted system, things will have the capability to cavort, coordinate, learn, and heal. A home and the people who live in it will begin to function more like a natural system—constantly monitoring all of its components and functions, coordinating activities that go on there, continuously learning and improving and adapting, so that the home is a place where technology exalts human beings and enables them to realize their needs and desires in the most complete, seamless, and satisfying way possible.

Enchanted systems also have the potential to take us into terra incognita, unknown and uncharted technological and human territory. Picture a world in which *every* object connects, has an ability or two of enchantment, will be embedded with sensors and signaling, and will climb a couple of rungs on the Ladder of Enchantment.

Such enchanted systems will enable a kind of superrich, interconnected intelligence and a network of one-to-one, one-to-group, and group-to-group communications of such speed, capability, and reach that they will make our current plex of phones and social media look rinky-dink.

I do not suggest that we will gradually develop one central brain or that human and machine will eventually meld into a long-lived, ultracapable, snap-in-new-parts hybrid. My vision is not of the all-knowing, creepy (and ultimately devious) HAL computer from *2001: A Space Odyssey*. I see something closer to a high-level version of the termite mounds and beehives that MIT's Mitch Resnick describes in *Turtles, Termites, and Traffic Jams: Explorations in Massively Parallel Microworlds*. These are complex, self-regulating systems that operate with no central brain at all. They respond to simple rules of organization (along with chemical and electrical signals) that all members of the community know and instinctively follow. The result is a resilient system.

Enchanted systems will operate in much the same decentralized way. The sensors and signaling will not respond to one fragile, totalitarian computer controlling billions of minion objects. Rather, an enchanted system will operate according to principles that will look more like a natural ecosystem than a computer network: self-regulation through feedback loops, resilience, adaptability, and—the great difference from natural ecosystems, which don't take human beings into account—respect for human desires and human concerns at their core.

Creating and managing these systems will bring new challenges to designers, technologies, product makers, as well as community and organization leaders. As MIT's Jay Forrester, the progenitor of systems thinking and author of the seminal *Industrial Dynamics*, adumbrated long ago, the dynamics of systems yield surprises. What makes sense as a way to improve things may actually make them much worse. Forrester's classic case: building too much low-cost housing in New York, aimed at eradicating poverty, actually made it worse, by displacing job-creating employers.

Similarly, the creation of enchanted systems is likely to lead to unintended consequences, so—for people in the enchantment business —knowledge of systems science will eclipse even computer science in importance. We must recruit experts in systems dynamics to augment the already diverse enchanted-object teams at universities and companies—people who understand feedback loops, positive and negative, and who can successfully operate in an environment that can be unpredictable and changes constantly.

If we can face those challenges with open eyes, equipped with relevant skills, sufficient knowledge, and appropriate skepticism and wariness—along with optimism and goodwill—we can look to a future world characterized by man-made enchantment that rivals that of nature.

In this final part, I will consider objects and services that are in our environment now, and ones in academic and industry research labs, that give us a sense of how the proximate future of enchanted systems may look.

I will explore three systems: homes, workplaces, and cities. What objects will we find in each? How might these objects collude on our behalf? What old and new human desires will they fulfill? What are the signals and feedback loops? What unintended consequences can we project?

TRANSFORMER HOMES

HOME SHOULD BE an enchanted place, but often a house or an apartment or a yurt is far from that—cramped, isolated, dysfunctional, expensive, repair-needy. It can take years of living in, taking care of, and customizing a dwelling before it starts to deserve the sobriquet *home*.

We have talked about a number of home-based enchanted objects—clocks and doorbells and picture frames and furniture—but have not addressed the physical space itself. How can the abilities of enchantment—particularly affordability, usability, and loveability—be infused into six hundred or two thousand square feet of living room?

CITYHOME:
PROTEUS IN THE LIVING ROOM

Let me introduce Mark. He is in his early twenties, with a fresh college diploma in his backpack, and has just taken his first job in the city. The job is good and pays decently, but the money is hardly on the home-buying level, and even renting an apartment is a stretch for Mark right now. So he does the obvious: moves back into his parents' suburban home. Most days he spends an hour and a half commuting.

Mark hates paying for gas, despises inching along in city traffic, and feels guilty about his large carbon footprint. Living at home isn't ideal for his social life either, but he doesn't think he can afford to move out and into a place of his own.

Mark needs CityHome, a new microapartment invented by a team at the Media Lab that offers transformer technology and home-scale enchantment. CityHome is an incredibly efficient, and therefore inexpensive, fifteen-by-fifteen-foot space that automatically transforms to accommodate everything Mark needs to do in his personal space: study, exercise, lounge, entertain, eat, and sleep. All without compromise.

The CityHome is controlled by this tangible clock. Just slide the activities around to set their start and stop times, or tap an activity to transform the apartment into that mode.

Mark controls the home with a wall-mounted device that looks like one of our most everyday of objects—a clock with concentric circles. Each activity is represented on the clock by an arc, and by configuring the arcs, almost like the hands of an alethiometer, Mark creates a routine for his day.

In the morning, once out of bed, his room moves into exercise mode: the bed lifts away into the ceiling, the floor space clears, and a full-wall, live video projection of a yoga studio starts with an attractive, motivational instructor. Mark is happy to spend thirty minutes, as the sun rises, doing his yoga with other virtual tree-posers. As he showers and gets dressed, the apartment hastens into kitchen mode. His kitchen wall opens and an island swings down from its nest in the wall; images of vegetarian omelets and fruit smoothies are projected onto the walls to encourage him toward a healthy breakfast.

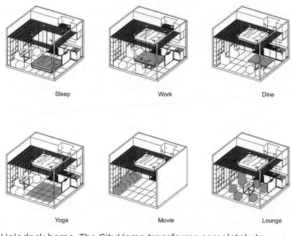

Sleep   Work   Dine

Yoga   Movie   Lounge

Holodeck home. The CityHome transforms completely to accommodate six activity modes. The bed is lifted and nests into the ceiling, and the dining table disappears into a thickened wall.

After eating, his desk descends from the ceiling, the overhead task lighting glows more intensely, window blinds close slightly to shield his screens from sunlight, and a couple of coworkers in India, just finishing their workday, are projected onto the far wall. Their collaboration Smart Boards are projected onto Mark's wall to help him understand their recent progress and where they need his help. After a couple of hours of focused work, he leaves the apartment for face-to-face meetings, which are within walking distance. (He grabs his Ambient umbrella, gently pulsing blue to indicate a rain shower is approaching.)

While Mark is away, the apartment cleans itself and then resets the furniture nesting into the walls and the ceiling. In the evening, when it senses his return with a couple of friends in tow, the furniture sets itself into lounge mode: chairs rise from the floor in the corners of the room, pinspot lights illuminate a little cocktail table that rises out of the floor, colored wall washes undulate like the inside of a lava lamp, music rises. The dining table can accommodate Mark alone or extend to banquet mode.

You can think of the home as a structural version of the shape-changing Greek sea god Proteus, who could transform himself into almost anything: a tree, a pig, a leopard. (Proteus was also omniscient.)

The inspiration for CityHome came from work at the Changing Places group at the Media Lab, where I participated in a semester-long charrette (an intense, collaborative process of design development) that focused on the housing challenges in the city of Shanghai. Design thinking is urgently needed for the rapid urbanization of countries such as China and India, as well as for the rapidly clogging city spaces of the developed countries. Urbanization is one of the great forces that will affect us all in the coming decades. Currently more than half the world's population inhabits city areas, and estimates are that well over 6 billion people will be living in cities by the middle of this century.[1] Asian cities are growing the fastest. The housing stock, infrastructure, and support services in these areas are often inadequate—so the more useful and flexible your apartment, the happier your life is likely to be.

Much more work needs to be done to commercialize and market transformable living spaces. China leads the world in understanding the urgency of this challenge: How does one enchant ever-smaller homes—and, as the middle class expands, the demand for high-tech goods and services grows? Beyond that, how can adding sensors, connecting the systems, tapping personal data, and getting objects in the system to work together bring a sense of enchantment?

MICROSPACES WITH MACRO COMMUNITIES

Another way to approach space constraints and create enchanted housing systems is to think differently about private and public usage. Does every person need a dedicated living room, exercise area, kitchen, study, and guest room?

During my senior year in college, I lived at the 92nd Street Y in

New York City for a term. My microroom was barely large enough to fit a bunk bed, and the only activities feasible there were those I could accomplish within the bed or the narrow strip of adjacent floor. All the other things I did that term—eat, talk, study, think, read, exercise—I did in communal spaces: cafés, pubs, libraries, parks, streets. I was a college student without a spouse or kids, so my space needs were individual, limited, and controllable. I loved that lifestyle, and although I now have a family, I still conduct many activities—including the research and writing of this book—in public areas.

The many models of shared living spaces offer individuals sufficient private space but shared common areas to reap the benefits of a community. Models of microspaces, too, are everywhere: college dorms, trains, boats, planes, campers, tents, yurts, and the "tiny house" movement—entire homes, with most of the traditional amenities, cleverly designed into a couple hundred square feet or portable abodes as small as seventy-three square feet. Many of these spaces serve as temporary or vacation homes.

In a country where big has always meant better, a new age is slowly dawning where small is beautiful and desirable. It's intriguing to note that small spaces, from houses to yachts, have always appealed to people with the financial means to afford big things. Millionaires adore their trekking tents and billionaires get away to their rustic lakeside cabins. Okay, Bill Gates's cabin is fifteen thousand square feet and cost $9 million to build,[2] but it does have a cabinesque feel, compared to his sixty-six-thousand-square-foot, $63 million primary residence on Lake Washington.

The lesson is that small spaces are great candidates for enchantment. We learned this lesson with cars, too. In the United States, it used to be that the only candidates for automotive enchantment were big vehicles. Then we learned what European automakers and drivers have known for decades: small can be every bit as usable, lovable, and storied as can huge.

To get a better understanding of how small spaces could be enchanted, our CityHome team conducted a survey to learn more about how often we move between activities in our homes. We found that houses support six primary modes of activity, and we ranked them, a little bit tongue in cheek, by how mentally engaging the activities were, from highly generative to passive.

1. Generate (work)
2. Chop and chew (food prep and eating)
3. Shred it (exercise)
4. Lean back and lounge (casual socializing with others)
5. Entertain me (immersive movie experience)
6. Ease in and out of sleep (the essential afternoon catnap)

Our insight was that people aren't interested in space for the sake of space, with the exception of wealthy people for whom houses are primarily emblems of financial achievement. Generally, people seek environments that will help them fulfill one or more of the fundamental human desires: to make them feel safe, to help them live a long and healthy life, to nurture their need for creative expression.

Even Bill Gates's main house, although it is almost as big as the lake it nestles beside, has undeniable elements of enchantment. Like the CityHome apartments, the Gates home climbs to the second rung on the Ladder of Enchantment: personalization. All inhabitants, as well as the favored guests, are given a smart tag that enables them to control many of the features of the rooms they enter or inhabit, including the temperature, lighting level, and music selection, as well as the most talked-about feature of the house: the digital versions of the world's famed artworks. This specific, dedicated deployment of the black glass slab—for the display of art, no app icons in sight—is intriguing, with elements of enchantment. It is an everyday object, the picture frame, augmented. Easy to use. Ambient. Well done, Bill Gates, you *can* do simple!

Some of these ideas are informing products and services that are now in development, on the cusp of commercialization, or already in the market—if not for the realization of complete enchanted-home systems, at least in the form of components and subsystems that will steer us in that direction.

BANG & OLUFSEN: BEAUTIFUL,
TRANSPARENT, POLITE

When I was a kid, three stores along State Street in Madison, Wisconsin, captured my imagination: the bike store Yellow Jersey; the camera store, filled with lenses, lights, and darkroom chemicals; and the electronics store called the Happy Medium. I was such a frequent and tenacious visitor to the bike store, always rebuilding bikes for my family and friends, one day an employee yelled to the owner, "Hey, Andy, let's give this kid a job already!" So he did.

At the camera store I grilled the owner on the best setup for building my darkroom below the basement stairs with a built-in light table for reviewing negatives, and the spooky red light to view my enlargements before they went into the fixing bath.

But the stereo store was the most enticing, partly because I couldn't afford anything there, but mostly because of one product that took my breath away: a gorgeous tower of cool, smoked glass. When you came near, the glass parted at a nearly invisible seam and the two panels silently slid open to reveal the cassette tape player within. It was, of course, a Bang & Olufsen design. Nowhere else had I seen a product that would respond to a human presence. It was magical.

Years later, when I could finally afford it, I purchased the B&O Beo9000 system, a combination CD player, tuner, and preamplifier. Other brands offered component systems: stacks of sleek black boxes, their faces dotted with milled knobs, meters with bloodred needles, and tiny, blinking LEDs—quite beautiful in a cold, electronic, nonhumanistic way. The Beo9000, however, mounted to a wall or stood on a soaring polished-aluminum leg and boldly displayed six CDs face out, so you could read the printing on them, rather than sucking them into a slim, mouthlike slot, as other CD players did. The player mechanism would glide to the CD you'd selected, the CD would spin, light glowing at its edges, and music would burble forth. The CD player was enchanting: a functional system and a work of art, self-evident and transparent. Most of all, it fulfilled my desire for creative expression in a particularly austere, metallic, techno-sculptural way.

The critical point here is that Bang & Olufsen, in the creation of its products—which include audio systems and loudspeakers, televisions, phones, and other products—thinks in terms of systems and of human interactions. The company defines its ideal this way: "*Bang & Olufsen exists to move you with enduring magical experiences.*"

The components are designed to work together as a system and, as such, can make a contribution to the enchantment of a home. In a mini-version of a global Internet of Things, the B&O components understand and respond to each other. As you are playing music on your B&O player (whether you're Bill Gates, listening in your thousand-square-foot dining room, or Mark, in his CityHome lounge area) and the BeoCom phone rings, the music automatically lowers to allow talk. While the phone is active, a small wheel on the handset—whose default function is to enable you to scroll through your contacts—transforms itself into a volume controller for all the other B&O components in your home. In this way, the human need for communication takes precedence over the electronics in your life. With this relatively modest manifestation of an enchanted system, it becomes easier to imagine a home where all the objects and devices are aware of and responsive to each other and, above all, act in service to the humans who live there.

FARMING AT HOME

I said the enchanted home should be a place that helps fulfill the human desire for a long, healthy life, and one important element of that quest is food. It used to be that the home and its source of food were one (a farm) or in proximity (a farm or a market not far away). Today our food flies in from everywhere around the world and is purveyed in megamarts.

In response to the globalization of the production of our food, a number of efforts have arisen meant to bring its places of production and of consumption closer together, such as the locavore movement and the creation of community gardens in cities. Urban foraging is also on the rise—the harvesting of weeds and flowers and mushrooms, figs and stinging nettles, snails and even squirrels. Although I have not foraged myself, I have been amazed by the wide variety of edible and even delectable foodstuffs found in our parks and vacant lots, along our sidewalks and backyard fences.

Could we apply one of the principles of enchanted design—the augmentation of a familiar object—to our desire to eat healthily at home, even in a tiny urban dwelling? Jennifer Broutin Farah of the Media Lab's Changing Places group has a passion for reconnecting people to food systems. Her SproutsIO concept brings a farming system into your home or office to help bring sustainable eating practices into any built space, small or large.

Jennifer describes SproutsIO as "an interactive food cultivation system that enables everyday people to reliably produce and access healthy food in urban areas. The system reliably grows a wide range of tasty fruits and vegetables, from strawberries to bok choy."[3]

One of the ways your future home will take care of you is by growing healthy food to feed your family, right in your living space.

SproutsIO is a modular system, with stacked levels of growing chambers, or pods. The plants are fed with a nutrient mist delivered at the base of each growing pod, a completely soil-free method with 98 percent less water use and 60 percent less fertilizer needed in comparison to soil. The sensors and automated system can monitor growth, compare harvests, and collect data to compare one harvest to another and one type of crop to another—and all of this information can be accessed from your mobile device or shared across a network. What grows best? What grows fastest? Plants and vegetables generally grow faster aeroponically than they do in soil, so there can be more harvests per year—as many as six times within the same area.

Once individual homes become enchanted food-producing systems, the next step is to create enchanted neighborhoods and cities—all focused on healthy food production for their communities. Imagine a city of home farms, managed by home farmers, all of whom share practices with each other, track progress, encourage those who need it, and celebrate those who make exceptional contributions, operating as a massively interconnected and resourceful community-supported agriculture system. Need parsley? Check online to see if any is available in the system. Wondering if your harvest will be ready for guests who are arriving two weeks hence? Check the database on a similar harvest. Having trouble with blight on a vegetable or brown spots on your lettuce? See what SproutsIO analytics, generated by the sharing of sensor information across all farms, has learned about optimal settings for producing lush and luscious leaves.

The vision: a town, a city, an urban sprawl, that becomes a community garden—a wellspring of connection, healthiness, and safekeeping—with an invisible scaffold of supporting technology.

THERMOSTAT 2.0

Henry Dreyfuss, an industrial designer, developed the first-round Honeywell thermostat in 1941. The device was simple and easy to use. According to the description from a Cooper-Hewitt exhibition:

> By 1953, the thermostat was refined into its now-familiar form, referred to simply as the Round. Its low price and ability to fit most situations has made the Round one of Dreyfuss's most successful designs. His continuing emphasis on ease of use and maintenance, clarity in form and function, and concern for end-use helped make Honeywell a leader in the field of controls in both domestic and industrial environments.[4]

Yet today, the energy savings promised by programmable thermostats are being lost because of lack of usability. The thermostat is trending toward the black glass slab in complexity and frustration. The Dreyfuss design has largely been replaced by a rectilinear minicomputer, with a display screen, multivariate settings, hard-to-read readouts, menu-driven controls, and sometimes tiny buttons.

Here's how Alan Meier, a senior scientist at Lawrence Berkeley National Laboratory and codirector of the Energy Efficiency Center at the University of California, describes the problem in a 2010 article, "Thermostat Interface and Usability: A Survey":

As many as 50 percent of residential programmable thermostats are in permanent "hold" status. Other evaluations found that homes with programmable thermostats consumed more energy than those relying on manual thermostats. Occupants find thermostats cryptic and baffling to operate because manufacturers often rely on obscure, and sometimes even contradictory, terms, symbols, procedures, and icons. It appears that many people are unable to fully exploit even the basic features in today's programmable thermostats, such as setting heating and cooling schedules.[5]

Tony Fadell is out to change all that. Fadell is the designer (and now entrepreneur) who led the teams that brought the Apple iPod into the world and has some three hundred patents to his credit.[6] He had not spent a great deal of time thinking about thermostats until he began building a new home near Tahoe. It struck him that this ubiquitous electronic device (roughly 250 million units are installed in the United States) came in one of two varieties—the simple, traditional, dumb version or the overcomplicated, ugly glass slab, which was also dumb. With the cost of energy inexorably rising, and more and more people using their living spaces as workplaces, too, he saw an opportunity to reinvent the thermostat.

Fadell applied his guiding design principle—"If you don't have an emotionally engaging design for a device, no one will care about it"—to reinvent the HVAC-controlling device, developing Nest, the "learning thermostat." He returned to a pleasing aspect of the traditional thermostat, roundness. The readout is simple, showing temperature and status and timing. You can manually set the thermostat by turning it like a doorknob, and each time you do, the Nest takes note of the time and tucks the data into its algorithm. You can also control Nest via an app on your mobile phone. Bit by bit, Nest learns your habits and preferences, senses your presence, and dials the temperature up and down to suit you and your activities.

I installed two Nests, one in my family's Boston apartment, and the other at our lake house in New Hampshire. I found that Nest saved money—heating the house only when we were at home—and also acts like a climate attendant, with the capability to fine-tune the environment to your liking. For the lake house, I keep the heat as low as possible in the winter when we're not there. When we get in the car in Boston for the two-hour drive northward, I use the app to crank up the heat from its downtime forty-five degrees to sixty-eight. The house, tucked away in the snowbound woods, is cozy warm when we arrive.

This feature of Nest is more important than it sounds because it eases the transition for us, the human beings, from one environment to the next. Before installing Nest, we would approach the lake house in winter dreading the first two hours there, waiting for the heat to warm the frosty house, sometimes hurriedly building a fire in the fireplace so we could huddle before it for a few minutes as we got unpacked and settled in. A wonderful house is not so wonderful when it's cold. Now we can leave our Boston apartment—knowing that Nest will take it down to a cost-saving, environmentally responsible temp when we're away—spend two pleasant hours in the climate-controlled car, and dash through the frosty air into the warm environment of the lake house and instantly feel at ease.

The next step would be to broaden the home system even further, so Nest, car, phone, and GPS system are part of it—and all talking to each other. The car and Nest would coordinate, without my involvement, to adjust the temperature based on my location and rate of travel so the lake house reaches sixty-eight degrees the moment we open the door. Nest was acquired by Google in 2014 for an astonishing $3.2 billion in cash, precisely proving the value of elegantly designed enchanted objects.

Like Fadell's learning thermostat, other technologies and devices will increasingly adapt to humans. We will talk less and less about the necessity for humans to improve their computer literacy and focus instead on the importance for computers and technological objects to have "human literacy." Technology will learn us. We won't have to sit before our glass slabs pointing at icons or responding to prompts about what to do next. We won't have to instruct our tech devices with voice commands or gestural inputs. The thing will collect its data, refine its algorithms, make its inferences, and do the right thing.

COLLABORATIVE WORKPLACES

How DO OUR primal drives manifest at different scales, and in the context of work? Let's now focus on enchanted objects and environments for collaborating more efficiently, on coworkers and colleagues who operate at a distance, and on when teams get together in the same room.

COORDINATING TEAM COMMUNICATION

Companies around the world are aggressively encouraging employees to work differently, partly to improve the way people work and partly to cut the expenses of maintaining large facilities and paying for a great deal of travel. In the cost-cutting effort, two tactics—virtualization (working from home) and hoteling (dynamically assigning offices)—have proved effective.

People who work in virtual teams, with members scattered across time zones, find a number of frustrations. No matter how much they communicate through email, Skype, and the rest, they still feel disconnected. They find it difficult to know when people are "on work" (actually getting something done on a project) because they can't see if they are "at work."

To help virtual teams feel more connected and communicate more effectively, I worked with office-furniture maker Steelcase on the design of a prototype called the Team Tile System. This is a three-by-three grid of panels that fits into a Steelcase modular workspace system. Each panel is connected to a specific remote team member. It changes color and brightness to show the status of that person—on work, off work, quiet and not to be disturbed, or in conversation. (Not unlike the Weasley family clock) A similar prototype developed at MIT, the team garden, performs much the same function but with a shape-shifting display. It's a set of slim plastic "stalks"—one for each team member—set into a base. Blooming from the top of each stalk is the team member's business card or photo. When that person is available, the business card or photo lights up. When the person goes off-line or is otherwise unavailable, his or her stalk gradually wilts and fades over ten minutes. The design's use of biomimicry isn't just for fun: people find it less distracting than having to contend with the glow of another screen.

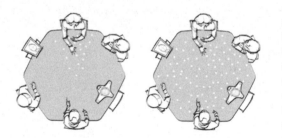

A constellation of lights gradually illuminates in front of each person as they speak. Over time, everyone can monitor the conversational balance of the discussion so introverts are encouraged and bullies are squelched. *Diagrams by Tellart*

SUPPORTING CONVERSATIONAL BALANCE

All businesses are interested in helping people collaborate more effectively, and enchanted objects have a role to play here, too.

As any meeting facilitator (or anyone with a tiny bit of emotional intelligence) understands, some people tend to dominate a conversation, while others need encouragement to contribute. Introverts have just as many good ideas as extroverts, but they can be reluctant to speak up, so their ideas are often suppressed. Their teams don't benefit from their ideas. Susan Cain's book, *Quiet*, inspired me to think more about this. She writes, "If we assume that quiet and loud people have roughly the same number of good (and bad) ideas, then we should worry if the louder and more forceful people always carry the day."

To help with this issue, I invented an enchanted table that makes this introvert-extrovert asymmetry visible on the surface of the table via subtle ambient feedback. Beneath the wood veneer of the table are hundreds of LEDs, which are initially hidden from view. As each person speaks, the LEDs directly in front of that person slowly illuminate—not all at once, but fading up over fifteen seconds—so they aren't distracting. After ten minutes or so, anyone can glance at the table surface and get a sense of the balance of the conversation. Who has spoken the most? Who has spoken less? Who has remained silent? In developing the table, I worked with a team of museum-exhibit designers at Tellart in Providence, Rhode Island, to get the brightness, pattern, and pace of illumination just right. If the LEDs change too fast or get too bright, it distracts from the conversation. If they change too slowly, the behavioral feedback loop isn't effective.

After testing the Balance Table in the lab with many meetings and casual conversations, retuning the parameters, and collecting comments from both introverts and extroverts after the test meetings, we found that people who use the Balance Table start to change in subtle ways. They avoid long soliloquies. Each person is more likely to encourage others to contribute, through smiles, nods, and body language. Once the tables were installed at Salesforce, a company whose culture is dominated by smart, often-introverted software engineers who were easily dominated by a lone loudmouth, we studied people's reactions again. These tables really work. They make people more attuned to one another. They encourage turn-taking and balance without intimidating anyone. People tend to prefer the conference rooms with these enchanted tables, and the subtle light of a lovely constellation of LEDs under the table surface feels a bit like the bioluminescent forest in the movie *Avatar*.

THE SCIENCE OF SERENDIPITY

In the 1940s, the Massachusetts Institute of Technology (as well as many other universities) constructed apartment buildings and other facilities to house student-veterans and their families. One such apartment complex, named Westgate West, would become the work site of three MIT professors—Leon Festinger, Stanley Schachter, and Kurt Back—and would fundamentally change long-held views about relationships, intimacy, and the importance of physical proximity.

Festinger and Schachter, both psychologists, and Back, a sociologist, based their study on Freudian psychoanalysis and the notion that someone's ability to build friendships (or not) asserted itself primarily through subconscious childhood experiences. Challenging this idea, Festinger, Schachter, and Back approached Westgate West several months after students had moved in, armed only with a floor plan of the building, and asked residents to list their three closest friends.

Forty-two percent of the time, residents named their next-door neighbor. Additionally, residents in apartments 1 and 5—situated at opposing ends of a hall, both at the bottom of a stairway—proved to be most popular in the building, if only because of their daily exposure to residents on the second floor. The findings, taken together, led the researchers to conclude that forming relationships and friendships was not necessarily due to having similar interests or attitudes, but rather was a matter of physical proximity.[1] The closer people were to one another, the closer they felt toward one another.

Cafeterias represent a special zone in the workplace. Though often dismissed as break rooms, places where people halt work, I believe the watercooler, coffee bar, and café provide one of the only reasons to drive to work in first place. Face-to-face conversation builds familiarity, rapport, and trust. Our progression from zero awareness of another person to full trust in the person happens faster when the interaction is lubricated by drinks and food.

So how can enchanted objects help with socialization? One way is with the architecture of long tables—think of the restaurants Waga-mama and Le Pain Quotidien, with their communal tables, and Hog-warts' huge dining tables—which encourage rubbing elbows with strangers and sparking serendipitous conversation. When I was at St. Olaf College, the cafeteria had rows and rows of long common tables, natural clique-busters, as the edges of one group inevitably brushed up against another.

The question is whether companies can get this same effect in a workplace cafeteria. And if your company is global, how can you encourage the same kind of interactions among and across sites without incurring onerous travel expenses?

Conversation Portal. People can share a lunch table across offices to encourage informal serendipitous interaction at work.

The conversation portal can make this happen. A portal is a high-definition videoconferencing system attached to the end of a long café table, adding to the physical table an equal amount of virtual table. When people sit at the real table, they see people who are seated at the "other end of the table" but are physically located at another facility. People on both ends of the table can pull up a chair and join in the conversation. The lighting and color of the tables are the same, which makes the table feel like a continuous whole, eliminating any feelings of "otherness." The "diagonal" conversation lowers the threshold for interaction. People are more inclined to start talking if they don't feel completely physically committed at first.

Coordinating the look and feel of the real and virtual tables is key. We learned a lot from Cisco's telepresence about how to make many people in multiple locations feel as if they were in one space. For telepresence, Cisco specifies the lighting, the table colors, even the exact Pantone shade of the wall paint to make videoconference attendees feel as if they were in the same space, even though they might be in different countries and time zones. The conversation portal represents the full magic of technology in spurring the kind of serendipitous connections that so often generate winning ideas. Introverts seem to respond to these techniques as much as extroverts do.

HUMAN-CENTERED CITIES

FOR DECADES, CITY planners and architects have sought to improve the livability of cities. How do we make them more human-centered and culturally vibrant? How do we reduce the traffic that divides and suffocates? What can we do to make streets more walkable? How do we foster more social interaction? How do we make a city safer through its own density? And perhaps most interesting, how do we leverage urban density to improve the aesthetics and attractiveness of the city itself?

In her iconic book *The Death and Life of Great American Cities*, Jane Jacobs makes this point: "A well-used city street is apt to be a safe street. A deserted street is apt to be unsafe."[1] What makes the street vital and safe are what she calls "eyes-on-the-street." A flow of people both makes the street attractive to others, because we like to see each other, and gives a feeling of safety in numbers.

The key, says Jacobs, is densification. People crisscrossing the streets, running errands, scudding along to myriad destinations. People shopping in stores, sitting in cafés. All of these activities, the constant presence of people who are both active and observant, make a city desirable and safe. But the idea that a certain density of people attracts other people to a city is one that "city planners and city architectural designers seem to find incomprehensible."[2] So as we design enchanted objects, we must remember this lesson: a walkable city is critical to promote street life, a feeling of community, the web of public safety, networks of peer-to-peer interaction that constitute everyday interaction, and a feeling that you are seen.

One of the challenges in creating an active city is the presence of cars. People need incentives and nudges for them to move away from cars as their primary transportation mode. "Human bodies and human metabolism are built for walking five hours a day," claims my esteemed cardiologist. The enchanted city will be built for helping us walk from home to work to socialize, to play, to meditate.

If designers lack for ideas about how to do this, they can tap those of city residents. Using an app akin to *SimCity*—and loaded with data provided by a shared database that everyone contributes to—residents will experiment with and "test-drive" future city scenarios. They will then offer their suggestions: *Charge more for using your car at peak traffic times. Give away fresh smoothies in the park on Fridays. Pay bike-sharing riders a bonus for riding more than one mile and see the consequences to ridership and health.*

Our future cities will be filled with wonder and magic. Good feedback loops will give us amazing efficiencies. The hyperconnected city will exist in an era of expansive possibilities. To be sure, it will also have its pitfalls, as we learn the complexities of enchanted-system dynamics. But we will experience enchantment in altogether new ways as we begin to see the city as both an expression of ourselves and an expression of an algorithm for enriched urban living.

CARS THAT BEHAVE LIKE HORSES

In old Westerns, a drunken cowboy would wobble out of the saloon and whistle for his horse. The steed would appear and dutifully transport the cowboy, asleep in the saddle, where he needed to go. Cars need to capture some of this home-on-the-range magic by operating themselves. One of the most important enchanted behaviors our cars could perform is self-parking—a concept that has the potential to radically alter the organization of city space.

Self-parking will mean that we won't have to share our cityscapes with parking lots, parking garages, and cars parked along every street and stuffed into alleyways. Today "in some US cities, parking lots cover more than a third of the land area, becoming the single most salient landscape feature of our built environment," Eran Ben-Joseph, a professor of urban planning at MIT, told the *New York Times*. "Five hundred million parking spaces in the country, occupying some 3,590 square miles, or an area larger than Delaware and Rhode Island combined."[3] Self-parking will mean you won't have to share the road with other people who are desperately hunting for a place to stow their cars. Studies show that 30 percent of city traffic stems from people driving around looking for parking spots.[4] Parking lots will become green space or high-rent retail or residential space. The sidewalk will be widened to provide an avenue for bikers, cruising along near the walkers.

The way to capture the Holy Grail of shared car use is through autonomous pickup and drop-off. An autovaleting car could park miles away from where you live or work—where the parking facility is doesn't matter because you'll never go there. As long as your car arrives curbside on time, you won't care where it cools its tires. All you will care about is that, fifteen minutes before the appointed time, you can summon your car with a simple flick of your wrist (provided you're wearing an iWatch).

As we saw with Google Chauffeur, self-driving vehicles not only promise new productivity when driving, they will also help us avoid accidents and make roads safer. Adding sensors to automobiles to prevent low-speed crashes could create economic value of as much as $50 billion per year by 2025.[5] Cars can also network with other cars to understand upcoming hazards miles ahead. New models already employ dozens of sensors to provide a 360-degree view without blind spots. They operate even better than professional drivers: They never nod off late at night, drive "under the influence," or take their eyes off the road when selecting music, eating, navigating, texting, or undertaking any other risky feat of multitasking.

One problem is how autonomous, self-parking cars will communicate with city pedestrians when no one is in the driver's seat. This issue was explored in an "expressive headlights" prototype developed at the Media Lab. Like eyes, the headlights blink and focus their gaze to indicate that the car sees the pedestrian, and that the pedestrian can cross the street with confidence that he or she will not be run down. Enchantment, indeed. One issue that still needs to be worked out: the more the car is automated, the more liability the carmaker will have. This may slow adoption of the autonomous car by some companies, but recent developments by Mercedes and other leading brands—which already refer to their cars as "semiautonomous"—suggest that the self-driving, self-parking, pedestrian-avoiding car is inevitable.

CITY-FRIENDLY VEHICLES

Even with new types of vehicles, we're not going to banish cars from the city anytime soon. So how can we alleviate the traffic, congestion, noise, and other impacts that cars have on a walkable and culturally rich urban environment? Reinventing the automobile holds the most promise for making cities more equitable, enjoyable, sustainable, and enchanted. The CityCar experimental project at MIT captures and crystallizes some incredibly powerful ideas that will change the cityscape and the experience of transportation.

The CityCar vision is remarkable in three ways. First, it rethinks the "DNA" of the automobile: power is provided by four electric motors housed inside each of the wheels, which shaves off the nose of the car and completely eliminates the transmission and drivetrain. The design allows entry and exit through the front of the car when the windshield lifts. This allows three CityCars to fit in the space of one standard parking place.[6]

Second, since the CityCar runs on batteries, you recharge the car as you would an electric toothbrush. All you have to do is roll the car over an inductive pad in your garage, at work, or in the city. If there is a power failure, the cars can act as a backup generator, sharing power with vital devices in the home. On hot summer days, the cars can even send power back to the grid, offsetting peak loads. In instances where fleets of electric cars connect to the grid, fleets of batteries can work together to prevent backup power plants—which may well be running on dirtier fuels—from coming online. Since high smog days usually occur in summer, this will benefit everyone.

Third, the CityCar lends itself to transport sharing. The small vehicle folds into an even smaller footprint for easier parking. You can load a few dozen folded CityCars into the equivalent of a vehicle vending machine. With a wave of a smartphone, customers release a clean, charged car. FIFO (first in, first out), permits max recharging time for each vehicle.

The most sophisticated idea for CityCar is the use of dynamic pricing to evenly distribute cars. I used Zipcar in Boston for years, which works on a round-trip model: you must drop off the car where you picked it up. A one-way model is much more useful for consumers, however, because they would rather pick up a car wherever they need it and drop it off when they're finished with it, wherever that is.

But for the company running the car-sharing operation (and similarly for bicycle-sharing companies), clustering is a problem. To redistribute the CityCars, large trucks would have to drive around at night, distributing the vehicles to suitable places for morning drivers.

A better option—made possible by connectivity—is to integrate the rental system with an incentive system. If you price the car depending on drop-off location, people will take care of redistribution on their own. If you are time-pressed or price insensitive, you'll take the closest car (or bike) and drop it off wherever you choose. When you have more time or want to save money (or even earn some), you'll walk farther to fetch the vehicle from a crowded node and leave it at one with more open spots. As if by "magic," the car, like the flying carpet of old, would appear in the right spot at the right time—all thanks to the connected system backed by the wonder of analytics on company servers.

AN ENCHANTED WHEEL
FOR THE TRADITIONAL BIKE

Another solution to the problem of car-clogged cities is to get people out of their cars altogether and onto bikes. Many people still consider bicycling an unrealistic or unappealing commuting option, and one reason is that it requires too much effort. People would rather not arrive at work sweaty and out of breath. A team of MIT engineers, led by Carlo Ratti and Christine Outram, set out to solve that problem with their invention, the Copenhagen Wheel, announced at the 2009 Copenhagen Conference on Climate Change.

The wheel, which can be fitted onto nearly any bicycle, has the capacity to store energy and release it again: while you brake, energy collects, and while you power uphill or through a headwind, it releases, giving you a boost (via an electric motor) as needed. The wheel also has a brake system, the Kinetic Energy Recovery System (KERS), to provide extra juice to stop your bike, if necessary. Additionally, the wheel lets riders record their speed and distance traveled, find their friends throughout the city, inspect air quality, and even be notified if their bike suddenly starts moving when they are not aboard.

The Italian Ministry of the Environment, in conjunction with Ducati Energia, has developed a prototype of the Copenhagen Wheel that should go into commercial production soon.[7] Though untested at scale, thousands of bikes equipped with the wheel would not just improve city life by increasing bike ridership, but also, by collecting (anonymous) data from the wheels, city planners could draw conclusions and give recommendations related to traffic and smog. The bikes could transmit data from around the city, acting like a roving sensory web whose reach, sophistication, and utility could only grow.

PERSUASIVE ELECTRIC VEHICLES

Though the desire to avoid exertion may dampen some people's enthusiasm for commuting by bike, that is not the only reason they would rather plop into their cars than climb onto a two-wheeler. An MIT team, part of the Persuasive Electric Vehicle (PEV) project, conducted an ethnographic study to understand why people do and don't ride bikes to work. The results were surprising.

Topping the list was not the overexertion, but rather safety. People fear being a lone biker on the street. When more people are biking nearby, people sense that cars will respect their space because they have more presence on the road.[8]

As a frequent bike commuter, I've felt this myself and often seek out companion commuters. The ride from my apartment in the Boston suburb of Brookline to the MIT campus in Cambridge is only three miles. For two of those miles, I can ride on a bike path that runs along the Charles River. I have a lovely panoramic view of the Boston cityscape across the water and, on the near bank, can peek into the riverside boathouses of the Boston University and MIT crew and sailing teams. When I have to ride on the street, however, there is no time for gazing. I must keep mindful every minute of whether passing cars will respect my space.

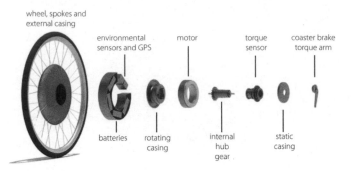

The most innovative feature here is embedding a motor, a battery, and wireless control in the existing envelope of a standard bike wheel. Now you don't have to worry about sweating up the hills during your commute.

How do we design a vehicle that will rival the car in its ability to efficiently deliver people to and from work—and keep riders of all kinds feeling safe? One of the goals of the PEV project at the Media Lab is to build a vehicle with a sense of enclosure and protection lacking in a standard bicycle. Riders will benefit from the increased safety provided by a small windshield, attached above their heads to their seat, and a sturdy back support. Open air on either side of the rider improves visibility compared to driving in a car and retains the easy in-and-out, off-and-on mobility of a bicycle. The PEV will also include an electric assist function, like that in the Copenhagen Wheel, which can store excess energy generated during travel. Its design will stress safety, all-weather travel, and ease of parking and recharging.

The team has taken behavioral economics into account, designing a system of incentives to get more people to pedal to work more frequently. One such incentive comes with the PEV's social-biking application, Spike, through which PEV riders can find PEV-riding friends throughout the city and ride in tandem with one another to receive special deals. The PEV and Spike will help lone bikers find other enthusiasts to ride alongside.[9]

SMARTER PUBLIC TRANSPORT

Smarter cars and better bikes will go a long way to making cities more accessible, cleaner, and less gridlocked, but they will not solve all the problems of getting around our conurbations. The cost of a shared car may still be prohibitive for some travelers. Journeys may be too long, or weather too inclement, for biking. We will still need public transport, but *enchanted* is probably the last word you would use to describe the bus, train, and subway systems currently operating in your area. How can we transform our public transportation systems to make them more human and attractive? How can we relieve the anxiety and increase the civility of using them?

Ken Schmier is a pioneer of improving public transportation and he believes that one important way to do so is to increase the information the system provides to its users. As the president and founder of San Francisco–based NextBus, Ken's mission is to make the experience of riding buses and trains less of a headache so that people will use the systems more and their cars less.

When Ken heard about my company Ambient Devices, he got in touch with me, wanting to discuss how our ideas about information display could improve public transportation. In a meeting with Ken that lasted an entire afternoon, I learned a lot about public transportation. First, Ken explained that the least expensive, most flexible, and most environmentally friendly form of transportation is a fully occupied bus. Second, he revealed the reason that keeps people from riding the bus more often. It's not the cost of the ticket, the presence of other riders, or the speed of the bus that discourages ridership. It's their aversion to uncertainty. When people are ready to go somewhere, they hate not knowing when and if the bus is going to show up. They can't stand the pain of waiting an indeterminate amount of time. If they don't know when the bus will arrive or have to wait too long, they will give up on it.

The bus operator faces a tremendous challenge in getting the timing between buses just right. Sometimes—as the result of traffic conditions, stoplights, and driving speeds—buses on the same route get too close together and arrive at the bus stops in clumps, with long waits in between. In the industry, the time between successive buses or trains is called headway. Studies show that if people must wait more than fifteen minutes—without getting any information about when the next bus will arrive—they're done. As Ken explained to me, "The city ends up driving empty buses around, at a huge cost, to keep a twelve-minute headway, so people will continue taking buses."

App-based transport systems are already on the market, but they don't work well for this type of information. They take too much effort to use even if the data is available. People tend to stop using them within a couple of weeks.

We found that people wanted to get the information about the bus at the bus stop. So we reinvented the familiar bus-stop pole, creating a standing beacon that connects to the bus company's GPS bus-tracking system, gathers real-time information about where buses are and when they will arrive at each stop, and displays it in a graduated fill-bar that can easily be seen from a distance. The enchanted bus pole relieves the anxiety that comes from a lack of information—*When is that damned bus going to show up?*—and enables people to plan their time better. They see they have, say, nine minutes until the bus arrives. They can do whatever they like, then casually saunter to the bus stop, and presto, the bus pulls in—like a ganglier, earthbound version of the flying carpet.

Not only do riders like the system, the city of San Francisco does, too. San Francisco spends about half a billion dollars to provide transportation to people in some 450,000 households, or about $1,000 per family. With the NextBus system, we found that by eliminating the uncertainty of when a bus will arrive, people become more patient—and they don't give up on the system even if the wait is longer than fifteen minutes. That means the bus company doesn't have to run so many mostly empty buses. This enchanted system changes the perception—and behavior—of an entire city of riders.

NET-ZERO BUILDINGS

In addition to more human-focused methods of transport, we need to enchant the buildings that front those happier streets of our future cities. Today, one the greenest buildings in the world is the Bullitt Center in Seattle. The center is a "net-zero" office building that generates all its own power and captures all of its water from rain. The building measures electrical consumption at every socket—metering power so closely that tenants get precise feedback to adjust their power use. And adjust they must, as they all agree in advance to work within a specified budget. Sensors monitor daylight and heat and automatically adjust window shutters to regulate light and temperature throughout the day based on room occupancy, weather, solar radiation, and more.

The sensors act as a nervous system, feeding data to remote servers that in turn create an environment for office workers with at least a small bit of fantasy: high-end, low-impact, delightfully climate-controlled, green, and responsible.

Such buildings are only a start. One challenge in cities is solving the utilization problem—or rather the woefully inefficient utilization. Consider hotels and offices. Most remain empty much of the time, essentially creating a double building footprint for every man, woman, and child using them. With fully connected buildings—using data gathered about how and when people use space, their schedules, and preferences—hotel rooms could turn into offices by day, offices into hotel rooms by night. We already see retail spaces doing double duty as restaurants, and classrooms at MIT live a double life as study spaces, choral practice halls, and party venues. Transformable spaces will help every building and room serve multiple needs, increasing the 24/7 utilization of precious urban space.

SHOPS ENHANCED BY MIRRORS WITH MEMORY

Even though we are buying more of everything online, retail stores will still play an important role in tomorrow's cityscape, and there will be myriad ways to enhance the shopping experience. Whatever your views on consumerism, cities without active retail businesses do not have the kind of densification that Jane Jacobs tells us is so vital to a city's health.

One of the most ubiquitous types of city retailer is the clothing shop. How can the inconvenient, frustrating, even maddening trying on of clothing be enchanted?

One answer is the mirror. People expect every clothing shop they visit to have mirrors, and mirrors are charged with meaning and potential for enchantment. In his book *Mirror, Mirror: A History of the Human Love Affair with Reflection*, author Mark Pendergrast describes the significance of mirrors over the millennia. In many ancient civilizations, the dead were buried with reflective metals or stones, mirrors were often associated with sun gods, and in the Middle Ages, people believed they could look into mirrorlike devices and see the future.[10]

The most relevant antecedent to today's retail mirror comes from mythology. Narcissus, an extremely attractive lad, gazed at his reflection in a pool and, unable to look away, died. This myth, and the notion that mirrors can show us an ideal identity, have inspired numerous cultural appropriations. In psychotherapy, for example, a treatment for pathological narcissism has been named the mirror transference.[11]

The most popular mirror-related fiction involves the queen in the Brothers Grimm's fairy tale *Snow White*. When asked by the queen who the fairest person is, the mirror always replies, "My queen, you are the fairest of the land." A startling real-life mirror tale comes from New Guinea. There, the Biami clan had traditionally not used mirrors. When at last they encountered them, the people began to groom themselves within a few days. During mourning in Judaism, sitting shivah, which means "seven," mirrors are removed from the scene to remove focus on ourselves. After a death, the grieving family shrouds the mirrors in the house and removes cushions from furniture for seven days as friends and family gather and receive visitors. Mirrors influence our sense of self—and self-consciousness.

The Mirror of Erised (*desire* spelled backward), from the Harry Potter series, shows its viewer his or her "deepest, most desperate desire."[12] Albus Dumbledore, the mirror's owner, comes across Harry sitting cross-legged in front of the mirror, entranced, and tells him, "The happiest man on earth would look into the mirror and see only himself."[13] Harry, an orphan, sees an image of himself with both parents by his side. Similarly, Dumbledore sees the reunion of an estranged family and himself brandishing a pair of thick, woolen socks.

As these experiences suggest, mirrors are almost magical in and of themselves. An augmented mirror, however, with interactive capability, could enable you to see an identity that you are actualizing. A social capacity could let you share this self with others. It could sate our hunger to see the future and our role in it. In the words of Mark Pendergrast, "The mirror appears throughout human history as a means of self-knowledge [and] self-delusion. . . . Mirrors ushered in the earliest human civilizations, and now they point us into the future."[14]

So, there is plenty of precedent for reinventing the ordinary mirror as a way to enhance the experience of city retail shopping. I'm working with an Israeli start-up company to do just that. The memory mirror, or MemoMi, is designed for use in retail stores, luxury hotel suites, and homes, anyplace where people are interested in trying on various outfits and getting feedback from remote friends about how they look. It can also be used to display images collected over time, to show a before-and-after change—for example, a difference in hair length at a salon or, at a gym, to show how your abs have become more defined.

When you step before the MemoMi mirror, it records a brief video of you. When you approach the mirror again, it displays the recorded image and the real-time image side by side. This will be a great boon to the clothing shopper. Choose outfit #1 and record how you look in the mirror. Try on outfit #2 and compare how you look in the two at the same time. If you have found more than two possibilities, no problem. Thumbnail images of the outfits stack up. Just point to the thumbnail you want, and you can compare it with what you're currently wearing. Of course there is a social function. You can access the images with an app and share the outfit options with your friends on Facebook or Twitter for immediate social shopping feedback. There's nothing like the wisdom of the crowd to help you determine your best look and also to encourage you to spend money on fashion.

SOLAR-POWERED WASTE MANAGEMENT

Another important factor in making cities more human-centered and livable is how we manage waste. Remember the smart trash can? Although the bin was able to affect eating, shopping, and recycling habits and, through story-ification, became a fond member of the household, it was still an isolated object. The trash can plays an even more important role when it joins an ecosystem of objects, comprising a fleet of public trash cans populating every city corner, gathering data, talking and working together, delighting city dwellers with cleaner streets and fewer garbage trucks.

Yes, lowly waste bins. And yet not so lowly. Consider the bins installed in a number of US cities by BigBelly Solar, a company based in Newton, Massachusetts. The bins are boxy, rodent-proof receptacles with built-in solar panels that power a trash compactor. They are placed on sidewalks and hold five times as much trash, compacted, as conventional trash cans can. They also have the ability to talk trash to the cloud about their status.

BigBelly trash cans embed connectivity and a solar trash compactor into the city's architecture to make pick up more efficient.

This capability may not seem like enchantment if you live in a wealthy area where the department of public works has a healthy budget that allows for regular trash pickup. But if you live where trash overflows its containers or attracts rodents, BigBelly cans become as storied as the home trash can. And if you're the head of the local public works department whose trucking and pickup time is cut in half—your trucks and haulers are automatically routed only to full cans—you're delighted to have a community of waste receptacles that talk to each other. BigBellies are also a favorite of people who care about the environment and want waste trucks to make fewer trips and burn less CO_2-emitting fuel. That's been the experience of people in cities from Boston to New York to Philadelphia.

BigBelly has only realized part of the potential of climbing the Ladder of Enchantment. The future of its trash bins might include socializing or gamifying recycling—making the bins an ever-more-engaging neighborhood appliance.

These examples give us a sense of how our way of understanding and living in a cityscape will change. We'll no longer inhabit an unanimated built landscape of homes, businesses, and parking lots. We will live within a web of connected and contributing objects: apartments, indoor gardens, parking meters, bus stops, urban furniture, signage, cars, and bikes.

Remember the movie *Avatar*? On the film's fictional planet, Pandora, all elements of the natural world can communicate with one another, and the world itself can communicate with Pandora's humanlike inhabitants, all of whom can link into the system through their individual biology, particularly their hair. It's not impossible to imagine human beings inhabiting Earth as a kind of modern Pandora, our fingerprints connecting us to every element of the planet's infrastructure.

FUTURE DRIVES

which can be satisfied by Enchanted Objects

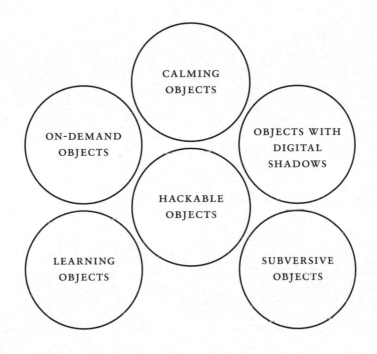

CALMING
OBJECTS

ON-DEMAND
OBJECTS

OBJECTS WITH
DIGITAL
SHADOWS

HACKABLE
OBJECTS

LEARNING
OBJECTS

SUBVERSIVE
OBJECTS

SIX FUTURE FANTASIES

THROUGHOUT THIS BOOK I have built on the notion that enchanted objects emerge from six perennial human fantasies or drives: for omniscience, telepathy, safekeeping, immortality, teleportation, and expression. But what new fantasies will surface in the future, and what objects or systems will satisfy them? I see six types of objects emerging to address new desires for our time. I say "emerging," but they are in a sense reemerging, as they reflect the universal desires but are distinct enough to warrant further attention. If you are an inventive company or business strategist, these trending themes may be the ones your innovation teams should focus on.

ON-DEMAND OBJECTS

Does your house have a guest bedroom that goes unused night after night? Last night forty thousand people rented a spare bedroom or couch from a stranger, thanks to Airbnb. A total of 2.5 million people used the service in 2012. The payment and reputation system is coordinated online. Now cars (RelayRides, Wheelz), beds, bikes, photo equipment, boats, outdoor gear, your driveway, and other things you own can be rented with remarkable ease or made available to rent from others—on demand, only when needed.

On-demand solves a problem that people everywhere face. Ownership of material assets is a never-ending hassle: you have to deal with storage, cleaning, upgrading, insuring, servicing, and paying fees and taxes. Most people would prefer to get the benefits of using things without these burdens. We want to have everything, yet own nothing. The solution, enabled by enchanted objects and cloud-based coordination, is sharing or "thin-slicing" ownership.

What other factors drive the desire to unburden ourselves and yet feel enriched as if we have immediate access to all the world's fantastic tools and resources? One is the age-old impulse to simplify life to its essentials—going Thoreau. Another is the innate pleasure we get from fully utilizing our assets. To both simplify and enrich our lives, we have to reduce what economists call transaction costs. These are the costs associated with finding, negotiating, pricing, rating, paying for things—all annoyances for customers in the nonenchanted world. With connected objects and cloud-based coordination, these costs and associated headaches vanish. We will be happier when we own less stuff but still have access to stuff when we want or need it.

On-demand sharing extends beyond hard goods and services. Three food-sharing websites are now operating in a limited number of cities. Feastly (eatfeastly.com) unites people who want to eat out cheaply with amateur chefs who want to cook and who volunteer to host (and feed) a group in their homes. Mealku lists meals prepared by one of a handful of participating chefs, available to be delivered to a user's door (within the hour). Super Marmite (Paris), like Mealku, lists available meals (as well as sides and snacks) and days, hours, and addresses where they can be picked up throughout the city. These food-focused services promise community, authentic experiences, and new tastes— all shared and not owned.

CALMING OBJECTS

Trends in food reveal another emerging fantasy: the desire for *slow*. Carlo Petrini of Bra, Italy, launched the Slow Food movement in 1986 to counter the advance of fast food—namely McDonald's—into his country. Petrini believed people should restore the connection between themselves and the producers of their food, and they should take the time to cherish food production, cooking, and dining. Today Petrini's movement has spread to 150 countries.[1] It inspired the "Slow Food Manifesto," released in a Paris theater in 1989: "Against the universal madness of the Fast Life, we need to choose the defense of tranquil material pleasure. Against those, and there are many of them, who confuse efficiency with frenzy, we propose the vaccine of a sufficient portion of assured sensual pleasure, to be practiced in slow and prolonged enjoyment. . . ."[2]

This desire for slow goes along with another universal impulse: the retreat to calm, quiet, and nature. Yes, we all love the richness and hustle and bustle of life, in cities and at cultural events. We like losing ourselves in social and artistic distractions. But today's beeping, buzzing world also creates a thirst for stillness to reflect, find ourselves, and create. (Remember Susan Cain's ideas about introverts and my table that helps them speak up.)

This impulse leads to a spate of new enchanted objects that restore stillness. The UK's Noise Abatement Society, a nonprofit organization dedicated to fighting noise pollution, introduced Quiet Mark, a seal-of-approval program, to certify objects fitting the trend. Lexus has tapped into this desire by playing up the quietness of its CT200h hybrid in a campaign dubbed "Quiet Revolution." In today's unquiet world, people increasingly want to be able to create quietude. We want stillness on demand, silence when we snap our fingers, and products and services that can do their work quietly and with great subtlety.

We see this drive reflected again in people's perennial desire to infuse nature—its calm and respite—into their cities, homes, and workplaces. We want to escape the pressures, pollution, noise, traffic, and other stressors of daily life. This desire partly feeds people's interest in home farming. It will also manifest in their embrace of organic elements in decor, biomimicry (using nature as a model) in design, and subtle, simple ambient displays.

Realizing the potential of enchanted objects to calm our environments will ultimately depend on the ecosystem of connected objects. These objects, working together, will neutralize the effects of behaviors that fragment and create noise in our lives. One of those behaviors is multitasking. The fetishization and glorification of multitasking is poorly grounded in the reality of how brains work. Switching back and forth between multiple tasks requires significant cognitive load. Studies of attention have shown that switching focus from one task to another wreaks havoc on productivity and concentration.[3]

But how do we bring calm and focus to our days when so many objects make demands on our attention? We need things that communicate with each other, cooperating to help us smoothly transfer our attention from one focus to another. To enable these transitions, we will have trusted filters or agents that understand our goals. A filter will have to operate so reliably that we feel comfortable *not* obsessively checking our email, vmail, and other channels. It requires analytics in the connected world to separate the trivial from the critical. With that kind of capability, the agent would then, with our blessing, transmit to our friends' agents a "Don't bother him unless it's critical" message.

The irony is that we have spent two decades developing and marketing technologies that invade and interrupt and support multitasking—to conduct three conversations at a time. My MacBook Pro has enough RAM to support about a dozen apps running simultaneously. But the universal thirst for full engagement in what we're doing—achieving a state of mind that University of Chicago psychologist Mihaly Csikszentmihalyi dubbed flow—remains unyielding. The unquenched fantasy is for a service to intelligently remove the distractions. Apps such as Freedom attempt to do just this, by turning off our always-on Wi-Fi connections so we can't check Twitter, Facebook, or email for a prescribed time. Once again, we are seeking to enrich life by subtracting elements from it.

HACKABLE OBJECTS

The material world has never been more programmable and more hackable than it is today. Two recent alumni of MIT's Media Lab have created magic in a drab-looking plastic square they call Twine. Though Twine looks like nothing more wonderful than a drink coaster, it encases a temperature sensor, accelerometer, and microprocessor and can communicate via Wi-Fi. It also has a plug-in for other sensors measuring moisture, vibration, and orientation. When conditions trigger one of the sensors—water flooding your basement floor, say—Twine sends a message via email, text, or phone to you, a handyman, or to some other connected device.

My students invented a simple motion-plus-temp-plus-humidity sensor that talks to the cloud called Twine. It can tweet, text, or email you based on simple "if this, then that" rules.

Twine delivers a great service—"Hey," an email warns, "the footing around your furnace is all wet!"—but it is also a harbinger of something else: a future of increasing hackability of objects. Twine provides a simple platform for that hacking. Not destructive hacking, mind you, but programmability, remixing, reassembly, and fast customization.

Since Twine can accommodate so many sensors and rules of communication, it provides a blank slate for programming your life. Need to tell house sitters that the cat litter needs changing? Make your desk lamp purr. Want to know when birds alight on your feeders? Have your phone chirp. Through the doorway of authorized hackability, people can create newfound enchantment—just for them. DIYH.

This points to the possibility of our realizing another persistent fantasy: to make our own magic. With the encouragement of Google, Apple, Microsoft, and hundreds of other companies, app developers can use public example code to build on and leverage other companies' software platforms and data streams. These bits of code are called APIs, or application programming interfaces. Programmers who want to combine their product's capabilities with another product's capabilities use the other company's API—and give customers the benefits of both products.

In a sense, all apps are mashups and recombinations. If a hotel wants to provide a customized map to give directions to customers from various locales, it uses the Google Maps API to bring in information from many online sources. The customer benefits from increasingly powerful, sophisticated, and easy-to-build experiences—the best of both software worlds through seamless sharing. The result could be an interactive map that shows not only directions, but also provides information on points of interest, gas stations, great restaurants, and clean restrooms along the way. APIs allow hackers to access and "play nice" with other companies' apps and data in a semicontrolled way. Both parties benefit, as does the consumer.

The era when only tech people do this kind of hacking is ending. Hacking the physical world will be as common as creating our own snap-together LEGO constructions. Not only geeks will combine and recombine data from various apps at their whim. Simple mashups are already within the reach of nontechies. Want to receive an email message every time the temperature in Houston hits one hundred degrees? Want to save any Facebook photo in which you are tagged to your Dropbox folder? Go to ifttt.com (If This Then That) and write a one-line rule today. Literally hundreds of services, all developed by different companies, can talk to each other. And the programming is easier than writing this sentence.

As APIs become available for every object and app, people will hack their way to using data and hackable objects in ways no one ever expected. This capability to use APIs to mix and remix will change our interaction with tangible objects. A celebration of this kind of hackability goes on display regularly at Maker Faires, amped-up shows for inventors and hobbyists. Inventors from all walks of life come together to show how they have combined and recombined technology to create new things. An example from the 2013 Maker Faire is a programmable illuminated vest. The vest, covered in 324 LED bulbs, can be programmed to flash different colors or patterns, and even to display text and video. Another good example is a robotic arm that allows teachers to reproduce any image (pictures, paintings, graphs) on their classroom's whiteboard.

Tangible hackability is coming to desks and kitchens around the world. Technologies such as 3-D printing suggest the possibility that we can create any object and "print" it at will. Want a desktop holder for a special souvenir from the beach or a new piece for a LEGO Mindstorms robot? Print it yourself. You don't even have to design it. You can download and tailor—or hack—designs from various Internet libraries.

Even wearable objects are hackable, and engineers are making them even more so. The Pebble watch, a wristwatch with an E Ink display, was introduced in the summer of 2013. Its makers provide an open API to allow others to write apps for the watch. It's just like a smartphone, except with a smaller, black-and-white screen. The seductive potential: anyone who buys a watch will design and program whatever service he or she desires. Control your home lighting. Lock or unlock your Lockitron doors. Change the temperature at the lake house. The watch, like the smartphone, offers a blank slate for inventive makers.

Like chefs with a cabinet of spices, we will all cook up enchanting works to suit our desires. We will become the magicians—because everything from shapes to behaviors will become hackable. And in the Internet of Things, each of the objects around us can talk with each other with those APIs. The vision is simple: the buyer of things will turn into the maker of things. In time, all matter becomes programmable—and programmable by everyone.

OBJECTS THAT LEARN

A vending machine is able to determine the gender and approximate age of a person who approaches it and then customizes its offerings based on a prediction of what the person will like. McDonald's is able to "predict what you're going to order based on the car you drive with 80 percent accuracy. . . . [Now] the fast-food chain [can] reduce the unacceptable thirty-second wait while your drive-in order is prepared."[4]

Cameras are becoming ubiquitous for security applications, but when given more computational power, and the ability to recognize objects, they can offer hints of the future. Imagine a security camera that recognizes and classifies suspicious activities, such as a person walking from car to car in a parking garage. Imagine your own camera recognizing the objects shown in the photo you take to predict what you would like to shop for, or to populate an online shopping cart with the contents of your friends' photos.

A large camera company showed me a prototype for a camera with an IKEA app that would automatically superimpose IKEA furniture onto a scene as you look through the camera lens. Don't like the furniture in the first photo? Swipe to the next. The furniture scales and rotates, the perspective in the scene changes, and colors subtly change as the direction of the light shifts. In just a few seconds, you could evaluate the suitability of chairs, lamps, and tables in your home, yard, or workplace. You can see how such an application could quickly predict, render, and persuade you to tap the "order now" button.

OBJECTS WITH DIGITAL SHADOWS

My fifth fantasy is the desire for my entire environment, not just objects, to act on my behalf. The fantasy is common in fairy tales: the woods come alive—every plant and animal and rock and river—to attend to your needs. This is not much beyond what's possible today. When connected objects talk to each other and work together on servers in the background, you will receive entirely new services. This is possible because each object will have a "digital shadow"—a parallel world of atoms and bits. Behind the tangible object is an intangible one in the virtual world with much more knowledge and connections.

To be sure, if you're a shopper today, you have already started to use your smartphone camera to keep track of and remember things you like—a friend's bag or shoes, a bike or a new car, furniture, gift ideas. We share these images via Facebook, Twitter, Tumblr, Pinterest, Path, Google+, Flickr, and Instagram with our friends and the general public. A social shopping revolution awaits. These images can be connected back to Amazon, eBay, and other retailers to allow people to shop through each other's photos. Think of it like an enchanted camera. You shoot, share, and purchase without ever entering a store.

My current company, Ditto, is built around the notion that we are inspired and influenced by our friends to decide where to dine, vacation, shop, or play. Social mimicry rules most of our decisions. If you see a car, just take a picture, and Ditto will superimpose a hot spot on the photo that links to an auction selling that car, right now, on eBay. Click another hot spot on the photo to schedule a test-drive. The retail world calls this distributed commerce. Snap a photo of anything: a shoe, a watch, a chair; the product is delivered that day to your door.

Ditto realizes a dream I've had about photography and the future of cameras. Imagine being able to click on any object in a photo to learn more, book travel, see prices, even buy it for a friend or yourself. Ditto has hundreds of servers in the cloud with image recognition software that can identify the contents of photographs. Any product in the frame is now clickable—shoes link to Zappos, people to LinkedIn, furniture to eBay, movie posters to Fandango tickets, restaurants to OpenTable reservations, beaches to Expedia travel deals. Every shared photo becomes an opportunity to act on the advice of your friends—the experience they are sharing. The camera becomes the enchanted object for learning and shopping—just point the camera and learn more about anything in the frame or even buy it from Amazon with a simple tap.

Digital shadows will enable services associated with every real object. You will be able to test-drive your taste in clothing on passersby on the street. They become like living mannequins, wearing items you are thinking about buying. Or walk into a hotel lobby and buy the furniture you see there, in the fabric of your choice, or browse through paintings available from the artists whose work is hanging on the walls.

Imagine every physical object now augmented with digital shadows. Information about these cameras at BestBuy is projected adjacent to them using a pico projector and a camera embedded in a lightbulb.

Wearable HUD displays make the digital shadow metaphor even more pervasive. Like a superhero, you can see through walls. To project your customized vision on reality. To selectively hide or highlight objects in a scene. To see in one place what happened there minutes before. With digital shadows, you will learn more, test ideas more quickly, gain deeper knowledge of any given situation. Imagine a plumber seeing through your wall—instead of tearing out a swath of gypsum board—to find where your water pipes are located. My students made a flashlight to do exactly this. Rather than just illuminating surfaces, the flashlight contains a GPS and gyroscope. It knows where it is and how you orient it. It also has a CAD model of the building, so it knows where the pipes and wiring are. Inside the flashlight is a data projector that can beam this information to display it on any wall. A mundane object in the real world, the flashlight becomes magical when it has a digital shadow.

ENCHANTED OBJECTS TO SUBVERT THE ENCHANTED OBJECTS

We see the precedents all around us of a final fantasy. There is the radar device in the hands of police, which is meant to protect us from reckless speeding drivers—but we spend hundreds of dollars on our own devices to evade that radar. TV and computer screens pop up in every location to entertain and educate us—but we thirst for a TV zapper, such as TV-B-Gone,[5] to silence them all. We carry cell phones that enable our anytime, anywhere communication, but we are desperate for a device to jam them when we're in theaters or concert halls, or bedrooms, or when driving.

Sometimes enough is enough—connected objects are leaning too hard on us from all sides. When they do, especially as they surround us in ecosystems, we may find that people who do not have our welfare in mind take them up in unintended ways. They may use them for devious purposes we dare not imagine. We cannot anticipate all the effects of these systems, and that brings us to the final fantasy: tech to foil overbearing tech, even those we willingly brought upon ourselves.

What might these tech foilers help us do? Block ads on the Internet and TV. Fool biometric security monitors. Evade public surveillance cameras. Confound analytics companies that load our computers with cookies and consolidate our data to send us tailored product offers. We feel vulnerable and fiercely desire to regain control of the controllers. Indeed, we hunger for higher-level objects to subvert the enchanted objects. Once we have tools for unlimited lifelogging and infinite memory, for example, we'll want services for therapeutic forgetting.

So this is the meta-wish: to thwart automation. To look inside our enchanted objects, understand their inner workings, and see that no magic actually exists. We want to open up the unfamiliar device or box to see what makes it tick. It's the cliché from so many military-industrial-alien-espionage movies. Operatives from a black ops agency arrive on the scene of some strange occurrence, surround it with tape, and shroud the bizarre beings that lie there in plastic. Then, wearing their superprotective suits, they approach whatever the creature is—E.T. or the Hulk or Superman or the Autobots—to reveal the secrets hidden within. Inevitably, our reductionist curiosity and micro-analysis kill the enchantment. As we hit the threshold where we realize we have ceded too much control, we'll want to wield magic neutralizing powers to regain it. Every technology needs a countertechnology to defeat it when it becomes too invasive. We will need a version of Harry Potter's invisibility cloak to afford ourselves a time-out from the enchanted ecosystem on steroids.

THE UNKNOWN UNKNOWN

These new fantasies—on-demand, calm, hackability, learning, digital shadows, and subversion—will lead us far beyond where we are today, to enchanted realms and ecosystems that impact how we live, learn, and interact with each other and the world. New home and city systems will become incorporated into existing national and global systems and structures. True, governments and despots will seek to centralize systems and threaten people's freedom and autonomy—dystopian visions abound. Yet a stronger force of entrepreneurs will spawn decentralized systems and enhance freedom, transparency, and self-determination. Not only will people tweet to assert their independence, the objects in the ecosystem will tweet, as well, and digital shadows from each object will make them easier to understand, configure, and use. Their data exhaust will help us detect helpful patterns and make recommendations that will enhance our lives.

Imagine a global system of connected objects in health care. Every diabetic might wear a sensor and a signaling system to track data on lifestyle, drinking and smoking habits, blood sugar, body weight, blood pressure, insulin injections, and medications. Each person would find the system enchanting, as colored orbs or smartphone apps or flowering "plants" conveyed reminders, warnings, and praise for medication compliance and blood-sugar control.

That would be just a start. For each community, the data would convey to physicians and health-care providers trends and warnings about diabetic populations and service demands. For drug companies, analytics would signal real-time results of medications, diet, and lifestyle changes—all grist for the development of better therapies. For device companies, data would show the performance and maintenance requirements for products such as insulin pumps. For global research labs, the data would provide crucial feedback on drug and device trials.

The globally connected system will astound, as people receive the daily hand-holding and guidance they might otherwise get only from a full-time nurse, an expert in diabetes management, equipped with an array of testing equipment. People in constant danger of putting their health at risk during a diabetic reaction could micromanage their diet and healthy behaviors to live the longest and healthiest life—and, to the delight of health-care companies, avoid costly trips to the ER. They might also tune into an app like Ginger.io—one that gives advice representing global intelligence tailored to their personal circumstances.

The diabetes system would be just one piece of an all-encompassing advisory system for every person with health concerns. Moreover, that system would fit into a mosaic of other systems—financial, work, entertainment—to analyze well-being. Any person who wants help living a better life, having more freedom, and achieving healthy functioning, financial security, and happiness— he or she will have it. This is the promise of technologically animating every object we deal with—and connecting those objects to talk and share and analyze and respond to our needs.

This is indeed terra incognita. But the New World of enchanted objects—the New Ecosystem, so to speak—promises new freedoms and new forms of prosperity. Old World ecosystems will pale in comparison. The risks of sailing to this new world remain. We need to heed warnings when we take some action that will veer us toward one of those dystopian visions. We must keep our eye on the opportunities, the benefits, and the unending delights that await us in this coming world—a world that exceeds the vision of enchantment we have been creating for ourselves, in fantasy and folklore and imagined technology, for thousands of years.

A METAPHOR
AND A MACRO TREND

I WILL LEAVE you with a useful metaphor, a macro trend, and a design approach from music.

First the *useful metaphor*: Think of the network as the new electricity. Connected products as the new electrification. Electricity is plentiful, invisible, and powers hundreds of products we take for granted. We rarely consider all those electrons running through every wall of our homes, schools, and businesses. Yet invisible as they may be, those electrons do flow, and we feel paralyzed during a power outage when the flow comes to a halt. Only then do we remember that candles and hand-cranked mixers and drills and phonographs were once the norm.

Try to think of the network as you (don't) think of electricity—as powering everything. Enchanted objects conversing with each other, all the time, through the cloud or peer-to-peer. You won't exactly know or care how—as long as the goal of this constant communication is for them to coordinate with each other and to serve you. The point is that everything will talk and share and gossip and scheme.

If you are in a tangible-product business, you need an enchantedobjects strategy. Around the year 1998, every company in the world figured out it needed to do something about the Internet. Remember how disruptive this was to every aspect of business—from marketing to product development to customer support and you name it? The disruptive power for the Internet of everything, for connected and emotionally engaging things, is even larger. Because now every object from pens to shoes, pill caps to furniture, to bikes and cars and even trash cans, will connect, engage customers in compelling new ways, and offer new services and business opportunities for those services that will completely change the game.

That's the metaphor. Now for the macro trend: *think swarms*.

Our fantasy for the future of artificial intelligence was partially delivered to us by Kubrick's *2001: A Space Odyssey*, in which HAL outsmarted the humans and left them powerless. How does HAL respond to Dave's request to open the pod-bay doors? "I'm sorry, Dave, I'm afraid I can't do that."

In broad strokes, what we have witnessed in the last twenty years is the atomization of artificial intelligence, AI. That is, we developed many single-function intelligences—multiple intelligences—working seamlessly together. Consider the modern car. No, we don't have KITT from *Knight Rider*. Instead we have antilock brakes, autodimming headlights, rain-sensor windshield wipers, a navigation system, cruise control—hundreds of technology piece-parts. Distributed.

The future of enchanted objects will follow this trajectory. We will have fewer devices with singular control, and more swarms of functionality. Not one smart thing, but a thousand just-smart-enough things, coordinated, working together.

Rather than having one bright light to illuminate a room, we will have a swarm of pinspots, able to shine and focus their light where it's needed. Rather than a large window-washing scaffold that progresses slowly up the skyscraper, we will have a swarm of small washbots that scour the windows when needed. Transportation gets smaller, too: think minivans rather than city buses. Even police work will take advantage of atomization. An eerie example is illustrated in the movie *Minority Report*, in which a small army of robots is able to scout a building with incredible efficiency, slipping under doors and scurrying through every room.

Swarms are more efficient, more fault tolerant, and more robust than single large devices, and it's the future of enchanted objects and functionality in general. Don't design a HAL for your business; design a hive.

Finally, *a design approach from music.* Let's call it sequencing.

Eliel Saarinen, the great Finnish architect, a pioneer of the art nouveau style, has often been quoted as saying, "Always design a thing by considering it in its next-larger context—a chair in a room, a room in a house, a house in an environment, an environment in a city plan."

In this book I have argued for the design of humanistic computer interaction through embedding functionality in everyday things. The examples range in size from pills and pens, to cars, homes, and cities. It's a power-of-ten approach in terms of scale.

Here's another way to think about how we'd like to interact with enchanted objects—an idea based on music. Music is played by following a score, a written document that defines what notes to play in what key and time signature. Notes are grouped into measures, measures into phrases, and further organized into themes and movements. Our lives often have the same sequence and rhythm. As my choir director Brian Jones reminds us weekly, "This goes to here, and then to there, and then to there." Every note exists within a phrase and should always be rising or falling in dynamics. Crescendo then decrescendo within a note, a phrase, and a song. "Pear-shaped tones, everyone! Don't let any note just sit there!"

This musical way of thinking can be applied to how we design the coming world of enchanted objects. Every interaction exists in terms of phrases and movements.

I hope my technophilic enthusiasm for enchanted objects is balanced by an understanding of the costs and losses that are inevitable with any societal change. I lament the loss of the front-porch culture that died when the automobile came along, and of the formal living room to receive visitors that became obsolete when the telephone took over our lives.

Enchanted objects are such a sweeping concept that it is difficult to imagine all the possible ways in which connecting everything will affect business and society.

Computer giant Cisco predicts between 50 billion and 1 trillion devices will be connected to the Internet in this decade, resulting in $14.4 trillion of economic impact.[1] McKinsey & Company, the seemingly omniscient management consultancy, in an article in the *McKinsey Quarterly*, claims that the Internet of Things will "create new business models, improve business processes, and reduce costs and risks." The McKinsey piece notes the importance of remote monitoring in health care, control technologies to better manage power and water systems, and the use of sensors in both public and private organizations to improve and "optimize business processes."[2] The potential to improve how we build, manage, learn, play, care for ourselves, and connect is enormous.

As we saw in the development of *Guitar Hero* and the GlowCap, the team that conceives and builds an enchanted object is like an octopus—with many arms reaching into professional domains, including industrial design, electrical engineering, mechanical engineering, computer science, business strategy, service design, branding, behavioral economics, and more. The diversity of expertise across a team of fifteen or twenty creators can be a little mind-bending.

As we learn more about the abilities of humans to sense and respond to technology, and as we invent new technologies for sensing and new materials for signaling, the capacity for these teams to create new and ever-more-enchanting experiences will expand exponentially.

We are a long way from even glimpsing the outer limits of this universe. Those who create enchanted objects have great room for invention. Still, given the many moving parts from hardware to software to providing ongoing services, it is easy to fail in the design, development, manufacture, and marketing of would-be enchanted objects.

I believe that a dialogue among business strategists, product designers, and technologists is necessary. Together we can answer the most interesting human-computer interaction questions for this time: What new opportunities exist for network-enabled hardware products? How does the world of the physical meet the world of the digital, each leveraging its own strengths? How do we craft emotionally engaging, useful, and benevolent services that we want to fold into our lives, day after day? How should those services be priced, especially when they encourage predictable behavior change like conserving energy, reducing medical costs, consuming more movies/books/medication, taking public transportation more often, ordering groceries automatically, or saving for a child's education? What long-term strategic advantage exists for innovative companies that rapidly deploy enchanted objects into homes, businesses and cities?

We will all be learning the answers to these questions together.

Be in touch,

David

drose@media.mit.edu

ACKNOWLEDGMENTS

In his book *What Technology Wants,* Kevin Kelly debunks the solo-inventor myth. He makes the critical point that we are all formed by the ecosystems that we inhabit. This certainly holds true for me. My network has defined, inspired, fueled, and helped realize everything that I've done.

I have benefited from and been shaped by many people in classrooms, industry conferences, and grungy start-up spaces, labs, and boardrooms. My brain has been reshaped and rewired through these interactions.

First I acknowledge my teachers: Don Hunt at Madison West for making art cool; physics professor David Nitz at St. Olaf for not giving up on me in quantum mechanics; George Brackett at Harvard for showing me that the educational power of computing is in construction kits and simulations, not drill and kill; and Gloriana Davenport at MIT for taking me under her wing in the Interactive Cinema Group, helping me fall in love with documentary film and the opportunities it offers for self-assembly, elasticity and video hyperlinks.

For inspiration, I thank the late Seymour Papert for inventing the Logo programming language and his foundational work with LEGO. He helped me understand that the most stimulating learning environment is one in which you make, invent, and build things. This vision guides the structure of the Media Lab at MIT today. There, Hiroshi

Ishii inspired me to think about ambient interfaces and to start Ambient Devices. I continue to enjoy teaching and collaborating with him in the incredibly inventive Tangible Media Group. I am indebted to Bruno Bettelheim for his book *The Uses of Enchantment: The Meaning and Importance of Fairy Tales,* which inspired the title of this book and renewed my interest in going back to folklore and myth to think about the future.

Over the last twenty-five years I've had the privilege of jumping off six cliffs (that is, founding six start-up companies) with an extraordinary set of people, many of whom are featured in the book. There is a unique combination of thrill, trust, and stress involved with starting a new company. I acknowledge my cofounders and early employees for their hard work and their willingness to take the risk of commercializing new ideas and technologies.

Thanks to the brave crew who started Interactive Factory: Zane Vella, David Clark, Robert Odegard, Glenn Johanson, David Curtis, Joan DeCollibus, Jean Wallace, Laurianne Serra, Gracia Gimse, Allen Yen, Mark Pine, Jose Fosse, Yuri Sebata, Harlan McCanne, Cheryl Tivey, Lao Lorenson, Joe Berkovitz, Chris Farnham, Marla Capozzi, and all of the museums, educational publishers, and toy companies like LEGO with whom we collaborated.

For recognizing the power of online photo sharing before the emergence of affordable digital cameras and camera phones, I thank my too-early-adopter team at Opholio: Neil Mayle, Rebecca Brown, Betsy Egan, Dita Vyslouzil, Emily Gutheinz, and Doug Robinow.

For exploring the "long nose" of digital life at Viant, I thank my cross-disciplinary team at the Innovation Center who helped generate many industry-changing prototypes still ahead of their time: Bob Gett, Tim Andrews, Hani Asfour, Bren Bataclan, Kate Ehrlich, David Tames, Josh Wachman, and Katherine Koh.

My amazing team at Ambient Devices who figured out how to build from scratch an elegantly designed consumer-products company with a Chinese supply chain, nationwide wireless network, and cloud-based configuration tools. Thanks to Pritesh Gandhi, Ben Resner, Nabeel Hyatt, Ert Dredge, Chris McRobbie, Mike Mooney, Mark

Prince, Katrina Haff, Sanjay Vakil, Andrew Boch, Noah Feehan, Gerd Schmieta, Bonnie Hamje, Fazle Khan, Dan Bradley, Joey Fitts, Johanna Schlegel, Mark Leiter, Myron Kassaraba, Jofish Kaye, and David Yett. Thanks to Yves Behar for inviting me to collaborate on a Novartis project that inspired Vitality's GlowCaps.

At Vitality, a small but incredible team was able to reinvent medication packaging and dance with the pharmaceutical and pharmacy behemoths. I am especially grateful for my longtime friend and cofounder Josh Wachman as well as Jamie Biggar, Julia Kim, Meredith Lambert, Nora Snyderman, and our prescient investor, Patrick Soon-Shiong.

Now, at Ditto, I'm lucky to work with a masterly team of computer-vision and machine-learning experts, including Josh Wachman, Neil Mayle, Philip Romanik, and Amy Muntz, to reinvent advertising through the recommendations embedded in photos.

> *"Never underestimate the power of a few committed people to change the world. Indeed, it is the only thing that ever has."*
>
> —*Margaret Mead*

Each of these start-ups depended on investors who believed in the vision, in the team, and in me. My serial investor, Nicholas Negroponte, deserves my special thanks for his continued support and encouragement. Nicholas believed in my ideas long before any evidence of market acceptance and has been a true mentor for more than two decades.

To my amazing students over the years at Yale, Mass Art, Marlboro College, Harvard GSD, MIT, RISD, and the Copenhagen Interaction Design Institute: I have learned so much from our interactions and your projects. Every day I'm impressed to see you progress and make important contributions in the world.

The community at MIT has been especially influential and stimulating, especially Muriel Cooper, John Maeda, Patti Maes, Bill Mitchell, Kent Larsen, and Sandy Pentland. Your important work pushing the frontier of interfaces forward frames this book.

Clients offer important hard problems, contexts, funding, and constraints that drive so much of design. I want to acknowledge Todd Pierce, for adopting my ambient furniture project within Salesforce .com, and Paul Franzosa and Shaun Salzberg plus their team at Tellart, who implemented the finished pieces. I've also enjoyed serving as a fellow at the international architecture firm Gensler, whose people understand and anticipate the transformation of their practice as they continue to integrate technology ever more elegantly into architectural spaces. Thanks especially to Gervais Tompkin, Jordan Goldstein, Thabo Lenneiye, Paul Franzosa, Arlyn Vogelman, and the visionary David Gensler.

I'm indebted to a host of longtime collaborators and friends, including Scott Kirsner, for nurturing the entrepreneurial community in Boston and running the fantastic Nantucket conference; Matt Cottam and Mike Kuniavsky, who run the Sketching in Hardware conference; Jef Huang, who taught with me at Harvard's GSD; the brilliant information visualization team of Mark Schindler and Angela Shen-Hsieh; JB Labrune, Timo Arnall, and James Frosch, for our conversations about the social history of invention, enchantment, and the critical link to psychological drives; and Richard Bergin, for supplying timely McKinsey data about the growth of the Internet of Things.

Thanks are also due to companies and conferences for asking me to visit and share the *Enchanted Objects* story, especially LIFT, TEDx, IDEO, Sogeti, Orange Institute, Xerox PARC, and the Institute for the Future.

For his phenomenal work researching, editing, and polishing this manuscript, I thank John Butman and his team at Idea Platforms, Anna Weiss, Henry Butman, Bill Birchard, Mark Brown, John De Lancey, and Kate Aurigemma. For illustrating the section heads and periodic table of Enchanted Objects that accompany the book (and for being so delightful to work with), I tip my hat to the talented Chris McRobbie. I highly recommend my editor Paul Whitlatch at Scribner and my agent Todd Shuster at ZacharyShusterHarmsworth, who have been wonderful and savvy guides in the book-making process.

Last, I'm grateful to my family for their understanding, support, and love when I become engrossed with products and start-ups. Thank you to my sister Susan and my parents, Sarah and Jim, for always encouraging my sometimes-baffling technology optimism. And thank you Sharon, my lovely wife, who sensitizes me to the unconscious emotional forces that drive good product design—and all other important things in life.

NOTES

PROLOGUE: MY NIGHTMARE

1 Arthur C. Clarke, *Profiles of the Future: An Enquiry into the Limits of the Possible* (New York: Henry Holt, 1984), 21.

1. TERMINAL WORLD: THE DOMINATION OF GLASS SLABS

1 Jeff Hecht, "Quantum Dot Displays Make Your TV Brighter Than Ever," accessed November 6, 2013, http://www.newscientist.com/article /dn23591-quantum-dot-displays-make-your-tv-brighter-than-ever.html.

3. ANIMISM: LIVING WITH SOCIAL ROBOTS

1 David Segal, "This Man Is Not a Cyborg. Yet," *New York Times*, accessed November 5, 2013, http://www.nytimes.com/2013/06/02/ business/dmitry-itskov-and-the-avatar-quest.html?pagewanted=all.
2 Carol Pinchefsky, "Dmitry Itskov Wants to Live Forever. (He Wants You to Live Forever, Too.)," *Forbes*, accessed November 5, 2013, http://www .forbes.com/sites/carolpinchefsky/2013/06/18/dmitry-itskov-wants-to-live-forever-he-wants-you-to-live-forever-too/.
3 Segal, "This Man Is Not a Cyborg."
4 *OED Online* (Oxford University Press), s.v. "robot, *n*," accessed November 4, 2013, http://www.oed.com.
5 Aristotle, *Politics*, trans. Benjamin Jowett, accessed November 4, 2013, http://jim.com/arispol.htm.

6 Mark Elling Rosheim, *Leonardo's Lost Robots* (Berlin: Springer-Verlag, 2006), 69.

7 Ibid., 112.

8 Isaac Asimov, "Runaround," *Astounding Science Fiction* 29 (1) (1942): 94–103.

9 *OED Online* (Oxford University Press), s.v. "neoteny, *n*," accessed November 5, 2013, http://www.oed.com/.

10 http://www.stanford.edu/~nass/books.html, accessed November 5, 2013.

11 Jonathan Welsh, "Why Cars Got Angry," *Wall Street Journal*, accessed November 6, 2013, http://online.wsj.com/article/SB114195150869994250.html.

THE DIALECTIC INTERPLAY: FICTION AND INVENTION

1 L. Frank Baum, *The Complete Wizard of Oz Collection* (with active table of contents) (Bedford Park Books, June 21, 2010), Kindle ed., locations 616–18.

2 Ibid.

3 Aristotle, *De Anima* iii 10, 433a31-b1, accessed November 5, 2013, http://plato.stanford.edu/entries/aristotle-psychology/#8.

4 Thomas Hobbes, *Leviathan,* accessed November 5, 2013, http://oregon state.edu/instruct/phl302/texts/hobbes/leviathan-a.html#CHAPTER I.

5 http://www.simplypsychology.org/Hierarchyofneeds.jpg, accessed November 5, 2013.

6 Center for Applications of Personality Type, accessed November 5, 2013, www.capt.org/catalog/Personality-Assessment-MBTI1.htm.

7 James Michael, "Using the Meyers-Briggs Type Indicator as a Tool for Leadership Development? Apply with Caution," *Journal of Leadership and Organizational Studies*, accessed November 5, 2013, http://jlo.sage pub.com/content/10/1/68.short.

DRIVE #1. OMNISCIENCE: TO KNOW ALL

1 Geoffrey C. Bunn, "The Lie Detector, *Wonder Woman*, and Liberty: The Life and Work of William Moulton Marston," *History of the Human Sciences* (1997): 96.

2 "Marston Advises 3 L's for Success; 'Live, Love and Laugh' Offered by Psychologist as Recipe for Required Happiness," *New York Times*, November 11, 1937, accessed November 5, 2013, http://query.nytimes.com/mem /archive/pdf?res=F20C11F93859177A93C3A8178AD95F438385F9.

3 Michael L. Fleisher, *The Encyclopedia of Comic Book Heroes: Volume 2—Wonder Woman* (New York: Macmillan, 1976).

4   Les Daniels, *Wonder Woman: The Complete History* (San Francisco: Chronicle Books, 2004). A striking parallel exists with an ancient text, Aristophanes's *Lysistrata*, in which women deny sex to their husbands in an attempt to end war.

5   Fleisher, *Encyclopedia*.

6   William Moulton Marston, "Why 100,000,000 Americans Read Comics," *American Scholar* 13 (1) (1943), accessed November 5, 2013, via JSTOR.

7   Susan Silverman, "Compass, China, 220 BCE," accessed November 5, 2013, http://www.smith.edu/hsc/museum/ancient_inventions/compass2.html.

8   Philip Pullman, *The Golden Compass: His Dark Materials* (New York: Random House Children's Books, November 13, 2001), Kindle ed., locations 1095–97.

9   Ibid.

DRIVE #2. TELEPATHY: HUMAN-TO-HUMAN CONNECTIONS

1   http://www.militaryphotos.net/forums/showthread.php?144073-quot-Ears-quot-amp-quot-Radars-quot, accessed November 5, 2013.

2   Harry McCracken, "Dick Tracy's Watch: The Most Indestructible Meme in Journalism," *Time*, accessed November 6, 2013, http://techland.time.com/2013/02/11/dick-tracys-watch-the-most-indestructible-meme-in-tech-journalism/.

3   J. K. Rowling, *Harry Potter and the Goblet of Fire* (Book 4) (Pottermore Limited, March 3, 2012), Kindle ed., locations 2322–23.

4   Hiroshi Ishii, unpublished paper.

5   Sherry Turkle, *Alone Together: Why We Expect More from Technology and Less from Each Other* (New York: Basic Books, 2011), Kindle ed.

DRIVE #3. SAFEKEEPING: PROTECTION FROM ALL HARM

1   Ian Johnson-Smith, *American Science Fiction TV: Star Trek, Stargate, and Beyond* (Middletown, CT: Wesleyan University Press, 2005).

2   John S. C. Abbott, *David Crockett: His Life and Adventures* (New York: Dodd & Mead, 1875).

3   Victor Appleton, *Tom Swift and His Electric Rifle; or, Daring Adventures in Elephant Land* (Kindle ed., May 16, 2012), locations 150–61.

4   Bruce Weber, "Jack Cover, 88, Physicist Who Invented the Taser Stun Gun, Dies," *New York Times*, accessed November 6, 2013, http://www.nytimes.com/2009/02/16/us/16cover.html.

5   http://www.taser.com/products/law-enforcement/taser-x26-ecd.

6   Lars Bevanger, "Doubts as to CCTV Efficacy in Big Brother Britain," accessed November 6, 2013, http://www.dw.de/doubts-asto-cctv-efficacy-in-big-brother-britain/a-16461692; and Rob Hastings, "New HD CCTV Puts Human Rights at Risk," *Independent*, accessed November 6, 2013, http://www.independent.co.uk/news/uk/crime/new-hd-cctv-puts-human-rights-at-risk-8194844.html.

7   Bevanger, "Doubts."

8   World Health Organization, "Road Safety: Estimated Number of Road Traffic Deaths, 2010," accessed November 6, 2013, http://gamapserver.who.int/gho/interactive_charts/road_safety/road_traffic_deaths/atlas.html.

9   World Health Organization, "Global Status Report on Road Safety: Supporting a Decade of Action," 2013, accessed November 6, 2013, http://www.who.int/violence_injury_prevention/road_safety_status/2013/en/index.html.

10 Ibid.

11 Richard Jackson and Benjamin Franklin, "An Historical Review of the Constitution and Government of Pennsylvania: From Its Origin, So Far as Regards the Several Points of Controversy, Which Have, from Time to Time, Arisen Between the Several Governors of That Province, and Their Several Assemblies: Founded on Authentic Documents" (London: R. Griffiths, 1759).

12 Mark Cieslak, "How Can 1.2bn People Be Identified Quickly?" *BBC*, accessed November 6, 2013, http://news.bbc.co.uk/2/hi/programmes/click_online/9722871.stm.

13 Adam Greenfield, *Everyware: The Dawning Age of Ubiquitous Computing* (Berkeley, CA: New Riders, 2006), 146.

14 Marc Langheinrich, "Privacy Invasions in Ubiquitous Computing" (Swiss Federal Institute of Technology, 2002), accessed November 6, 2013, doi:10.1.1.6.6743&rep=rep1&type=pdf.

15 Kashmir Hill, "How Target Figured Out a Teen Girl Was Pregnant Before Her Father Did," *Forbes*, accessed November 6, 2013, http://www.forbes.com/sites/kashmirhill/2012/02/16/how-target-figured-out-a-teen-girl-was-pregnant-before-her-father-did/.

16 Langheinrich, "Privacy Invasions," 6.

17 Rob Van Kranenberg et al., "The Internet of Things" (paper presented at the 1st Berlin Symposium on Internet and Society, October 25–27, 2011), accessed November 6, 2013, http://www.theinternetofthings.eu/sites/default/files/%5Buser-name%5D/The%20Internet%20of%20Things.pdf.

DRIVE #4. IMMORTALITY: A LONG AND QUANTIFIED LIFE

1 Herodotus, *The History of Herodotus*, vol. 1, trans. G. C. Macaulay (Kindle ed., 2011).
2 Jack Zipes, ed., *Spells of Enchantment: The Wondrous Fairy Tales of Western Culture* (New York: Viking, 1981).
3 Stephen Cave, *Immortality: The Quest to Live Forever and How It Drives Civilization* (New York: Crown Publishers, 2012), Kindle ed.
4 Andrew Goldman, interviewer, "Ray Kurzweil Says We're Going to Live Forever," *New York Times*, January 25, 2013, accessed November 6, 2013, http://www.nytimes.com/2013/0½7/magazine/ray-kurzweil-says-were-going-to-live-forever.html.
5 "Fm-2030," *All Things Considered*, National Public Radio, July 7, 2011, accessed November 5, 2013, http://www.npr.org/templates/story/story.php?storyId=1076532.
6 Douglas Martin, "Futurist Known as FM-2030 Is Dead at 69," *New York Times*, July 11, 2000, accessed November 5, 2013, http://www.nytimes.com/2000/07/11/us/futurist-known-as-fm-2030-is-dead-at-69.html.
7 Cave, *Immortality*.
8 Jason Gilbert, "HAPIfork: Buzzing Fork Offers Ultimate First-World Solution to Overeating," *Huffington Post*, January 1, 2013, accessed November 6, 2013, http://www.huffingtonpost.com/2013/01/08/hapifork-buzzing-fork-solution-overeating n 2433222.html.

DRIVE #5. TELEPORTATION: FRICTION-FREE TRAVEL

1 Edith Hamilton, *Mythology* (New York: Little, Brown, 2012).
2 Pennina Barnett, "Rugs R Us (and Them): The Oriental Carpet as Sign and Text," *Third Text* (2008): 14.
3 Ibid.
4 Gregory Benford and the Editors of *Popular Mechanics*, *The Wonderful Future That Never Was: Flying Cars, Mail Delivery by Parachute, and Other Predictions from the Past* (New York: Hearst Books, 2012), 135–36.
5 Ibid., 137–38.
6 Ibid., 143.
7 Ibid., 147–49.
8 Ibid., 151.
9 Alex Nunez, "The World's First 4WD Limo Is a Volkswagen," *Autoblog*, August 9, 2006, accessed June 1, 2013, http://www.autoblog.com/2006/08/09/the-worlds-first-4wd-limo-is-a-volkswagen/.

10 *Sunshine*, directed by Danny Boyle (Hollywood, CA: Fox Searchlight Pictures, 2007).

11 Cover of *Popular Mechanics*, July 1957, fig. 2 (Web: June 1, 2013).

12 Angela Greiling Keane, "Self-Driving Car More Jetsons Than Reality for Google Designers," *Bloomberg*, February 6, 2013, accessed November 6, 2013, http://www.bloomberg.com/news/2013-02-06/self-driving-cars-more-jetsons-than-reality-for-google-designers.html.

13 Volvo Cars, "Volvo Car Group Demonstrates the Self-Parking Car," July 8, 2013, accessed November 6, 2013, http://www.youtube.com/watch?v=GIa1mWr1kNs.

14 Scott Cleland, "10 Questions for Google Chauffeur," *Precursor* (blog), October 11, 2010, accessed November 6, 2013, http://precursorblog.com/?q=content/10-questions-google-chauffeur.

DRIVE #6. EXPRESSION: THE DESIRE TO CREATE

1 Mitchel Resnick, "Rethinking Learning in the Digital Age" (*Media Lab*, Massachusetts Institute of Technology, n.d.), accessed November 7, 2013, http://llk.media.mit.edu/papers/mres-wef.pdf.

2 Edward Rothstein, "In a Child's Tiny Bricks, the Logic of Computers," *New York Times*, October 23, 1999, accessed November 5, 2013, http://www.nytimes.com/1999/10/23/arts/in-a-child-s-tiny-bricks-the-logic-of-computers.html?pagewanted=all&src=pm.

THE EXTRAORDINARY CAPABILITY OF HUMAN SENSES

1 Herman Chernoff, "The Use of Faces to Represent Points in k-Dimensional Space Graphically," *Journal of the American Statistical Association* 68 (342) (1973), accessed November 6, 2013, http://www.apprendre-en-ligne.net/mathematica/3.3/chernoff.pdf.

2 "Soundscapes: Ecological Peripheral Auditory Displays," accessed November 6, 2013, http://sonify.psych.gatech.edu/research/soundscapes/index.html.

3 Mercedes-Benz, accessed November 6, 2013, http://www5.mercedes-benz.com/en/innovation/a-fragrance-for-the-new-s-class-interior-perfume-scent-sabine-engelhardt/.

4 Interview conducted by the author.

TECHNOLOGY SENSORS AND ENCHANTED BRICOLAGE

1 http://www.adafruit.com/products/391.

NOTES

FIVE STEPS ON THE LADDER OF ENCHANTMENT

1 *Quantified Self* (blog), accessed November 6, 2013, http://quantifiedself
.com.

TRANSFORMER HOMES

1 World Health Organization, "Urban Population Growth," accessed
November 5, 2013, http://www.who.int/gho/urban_health/situation_
trends/urban_population_growth_text/en/.
2 Nicholas Carlson, "Bill Gates' New $9 Million Country Cabin," *Huf-
fington Post*, June 5, 2009, accessed November 6, 2013, http://www.huff
ingtonpost.com/nicholas-carlson/bill-gatess-new-9-million_b_211914
.html.
3 Jennifer Broutin Farah, "Seedpod," accessed November 6, 2013, http://
www.jenniferbroutin.com/Projects/SeedPod/.
4 Robert Flinchum, "The Round Thermostat" (Cooper Hewitt, National
Design Museum, October 11, 2013), accessed November 6, 2013, http://
www.cooperhewitt.org/object-of-the-day/2013/10/11/round-thermostat.
5 Alan Meier, "Thermostat Interface and Usability: A Survey" (Lawrence
Berkeley National Laboratory, 2011), accessed November 5, 2013,
http://escholarship.org/uc/item/59j3s1gk.
6 Tony Fadell, Nest bio, accessed November 6, 2013, http://nest.com/
about/.

COLLABORATIVE WORKPLACES

1 Leon Festinger, Stanley Schachter, and Kurt Back, *Social Pressures in
Informal Groups: A Study of Human Factors in Housing* (Palo Alto, CA:
Stanford University Press, 1983).

HUMAN-CENTERED CITIES

1 Jane Jacobs, *The Death and Life of Great American Cities* (New York:
Random House, 1989).
2 Ibid.
3 Michael Kimmelman, "Paved, but Still Alive," *New York Times*, January 6,
2012, accessed November 6, 2013, http://www.nytimes.com/2012/01/08
/arts/design/taking-parking-lots-seriously-as-public-spaces.html?page
wanted=all&_r=0.

4 Donald Shoup, "Cruising for Parking," *Access* 30 (Spring 2007), accessed November 6, 2013, http://shoup.bol.ucla.edu/CruisingForParking Access.pdf.

5 Which-50.com, "McKinsey Says Healthcare and Manufacturing Will Gain Most from the Internet of Things," accessed November 6, 2013, http://which-50.com/post/52610861004/mckinsey-says-health-care-and-manufacturing-will-gain.

6 *Changing Places* (blog, MIT), "CityCar Changing Places Group," video, accessed November 6, 2013, http://changingplaces.mit.edu/research/projects/54-citycar.

7 Jack Schofield, "MIT Re-invents the Wheel, for Bicycles," *Guardian*, December 15, 2009, accessed November 6, 2013, http://www.guardian.co.uk/technology/blog/2009/dec/15/mit-copenhagen-wheel-green-bike.

8 *Changing Places* (blog, MIT), "PEV: Persuasive Electric Vehicle," video, accessed November 6, 2013, http://cp.media.mit.edu/research/videos/89-pev-persuasive-electric-vehicle.

9 MITChangingPlaces, "PEV Persuasive Electric Vehicle Combined with Social Cycling Application Spike," YouTube, December 10, 2012, accessed November 6, 2013, http://www.youtube.com/watch?v=FQhkj7ctEzw.

10 Mark Pendergrast, *Mirror, Mirror: A History of the Human Love Affair with Reflection* (New York: Basic Books, 2003), Kindle ed.

11 Allen Siegel, *Heinz Kohut and the Psychology of the Self* (New York: Routledge, 1996), 66.

12 J. K. Rowling, *Harry Potter and the Sorcerer's Stone* (New York: Scholastic, 1999), 213.

13 Harreh Pottah, "Harry Potter—Mirror of Erised Scene 2 HQ," YouTube, June 28, 2011, clip from *Harry Potter and the Sorcerer's Stone* (Warner Brothers, 2001), accessed November 6, 2013, http://www.youtube.com/watch?v=Kn7cR_8_vAg.

14 Pendergrast, *Mirror, Mirror*.

SIX FUTURE FANTASIES

1 Slow Food website, accessed November 6, 2013, http://www.slowfood.com/.

2 Peter Popham, "Carlo Petrini: The Slow Food Gourmet Who Started a Revolution," *Independent*, December 10, 2009, accessed November 6, 2013, http://www.independent.co.uk/life-style/food-and-drink/features/carlo-petrini-the-slow-food-gourmet-who-started-a-revolution-1837223.html.

3 Christine Rosen, "The Myth of Multitasking," *New Atlantis*, Spring 2008, accessed November 6, 2013, http://faculty.winthrop.edu/hinera/

CRTW-Spring_2011/TheMythofMultitasking_Rosen.pdf. This article is a meta-summary of research/studies/thoughts on multitasking and its negative effects on attention, the economy, productivity, etc., citing as far back as Henry James.

4   Andy Goodman and Marco Righetto, "Why the Human Body Will Be the Next Computer Interface," *Fast Company*, accessed November 6, 2013, http://www.fastcodesign.com/1671960/why-the-human-body-will-be-the-next-computer-interface?utm_source=feedburner&utm_medium=feed&utm_campaign=Feed%3A+fastcompany%2Fhead lines+%28Fast+Company%29.

5   CornfieldElectronics.com, accessed November 6, 2013, http://cornfield electronics.com/tvbgone/tvbg.home.php.

A METAPHOR AND A MACRO TREND

1   Joseph Bradley, Joel Barbier, and Doug Handler, "Embracing the Internet of Everything to Capture Your Share of $14.4 Trillion," white paper (Cisco Systems, February 12, 2013), accessed November 6, 2013, http://www.cisco.com/web/about/ac79/docs/innov/IoE_Economy.pdf.

2   Michael Chui, Markus Löffler, and Roger Roberts, "The Internet of Things." *McKinsey Quarterly*, March 2010, accessed November 5, 2013, http://www.mckinsey.com/insights/high_tech_telecoms_internet/the_internet_of_things.

INDEX

Page numbers in *italics* refer to illustrations.

PHOTOGRAPH AND
ILLUSTRATION CREDITS

Umbrella illustration © Chris McRobbie (page 1); wood-working tools photograph appears courtesy of the author (page 2); Sifteo photograph appears courtesy of Dave Merrill (page 4); barometer photograph appears courtesy of the author (page 5); Jack Zipes cover photograph courtesy of Jack Zipes (page 9); "Four Futures" diagram (page 15) and "The Inevitable Issue with Androids" diagram (page 40) appear courtesy of the author; Autom photograph appears courtesy of Cory Kidd (page 44); Livescribe illustration © Chris McRobbie (page 47); wallet photograph appears courtesy of the author (page 48); Shine photograph appears courtesy of Sonny Vu (page 53); Vitality card photograph appears courtesy of the author (page 53); "Six Human Drives" diagram appears courtesy of the author (page 59); Nike shoe illustration © Chris McRobbie (page 61); MemoryMirror photograph courtesy of MemoMi (page 63); "Omniscience" diagram appears courtesy of the author (page 69); Ambient Orb illustration © Chris McRobbie (page 71); David Rose and Ambient Orb photograph appears courtesy of the author (page 75); sketches for shape-shifting illustration appear courtesy of the author (page 77); energy clock illustration © Chris McRobbie (page 79); "Telepathy" diagram appears courtesy of the author (page 83); Home illustration © Chris McRobbie (page 85); Like-A-Hug photograph appears courtesy of Hiroshi (page 89); "Doorbell" diagram © Chris McRobbie (page 91); "The Value of Presence" graph appears courtesy of the author (page 93); LumiTouch photograph appears courtesy of Hiroshi (page 94); "Safe-keeping" diagram appears courtesy of the author (page 97); Tagg illustration © Chris McRobbie (page 99); onesie photograph appears courtesy of Mimo (page 104); wireless tag photograph appears courtesy of Tile (page 110); "Immortality" diagram appears courtesy of the author (page 111); HAPIfork illustration © Chris McRobbie (page 113); Beam toothbrush photograph appears courtesy of Beam (page 117); Glowcap illustration © Chris McRobbie (page 127); "Designing for Subtlety" diagram appears courtesy of the author (page 129); "Teleportation" diagram appears courtesy of the author (page 133); Terrafugia illustration © Chris McRobbie (page 135); "Expression" diagram appears courtesy of the author (page 141); Io Brush illustration © Chris McRobbie (page 143); *Guitar Hero* photograph appears courtesy of Harmonix (page 148);

LEGO brick photograph appears courtesy of ifactory (page 153); LEGO illustration © Chris McRobbie (page 167); "Seven Abilities of Enchanted Object" diagram appears courtesy of the author (page 171); Joule illustration © Chris McRobbie (page 173); bus pole illustration © Chris McRobbie (page 174); preattentive examples illustration appears courtesy of the author (page 175); "Ambient Displays Respect Your Attention" diagram appears courtesy of the author (page 176); Energy Joule photograph appears courtesy of the author (page 180); Amazon Trash Can illustration © Chris McRobbie (page 181); image furniture illustration © Chris McRobbie (page 182); ring illustration © Chris McRobbie (page 185); SunSprite pedometer photograph appears courtesy of Ed Likovich (page 185); Narrative clip photograph appears courtesy of the author (page 187); Facebook Coffee Table illustration © Chris McRobbie (page 188); musical chair illustration © Chris McRobbie (page 189); Nabaztag illustration © Chris McRobbie (page 190); cute Glowcaps illustration © Chris McRobbie (page 191); Withings scale illustration © Chris McRobbie (page 193); "Ladder of Enchantment" diagram courtesy of the author (page 194); customer feedback illustration © Chris McRobbie (page 209); CityHome controller illustration © Chris McRobbie (page 214); CityHome layout illustration © Chris McRobbie (page 215); SproutsIO photograph appears courtesy of Jennifer Broutin Farah (page 221); Nest illustration © Chris McRobbie (page 223); balance table illustration © Chris McRobbie (page 227); "Balance Table" diagram courtesy of Matt Cottam (page 228); Conversation Portal photograph appears courtesy of Matt Cottam (page 231); bike illustration © Chris McRobbie (page 233); Persuasive Electric Vehicle illustration © Chris McRobbie (page 239); wheel component diagram appears courtesy of SuperPedestrian (page 240); bus stop illustration © Chris McRobbie (page 241); BigBelly trash can photograph appears courtesy of the author (page 247); "Future Drives" diagram appears courtesy of the author (page 249); Twine photograph appears courtesy of Twine (page 255); wearable HUD displays photograph appears courtesy of Natan (page 260); guitar illustration © Chris McRobbie (page 265).

PHOTO INSERT

Page 1: (*top*) Ambient Orb photograph appears courtesy of the author; (*bottom*) LEGO photograph used by permission of LEGO. Page 2: (*top*) sketches by Tool for Ambient Devices; (*bottom*) dashboard photograph used by permission of Ambient Devices. Page 3: 5-Day Weather Forecaster photograph used by permission of Ambient Devices; (*bottom*) Energy Joule photograph used by permission of Ambient Devices. Page 4: (*top*) Glowcap with hub photograph courtesy of the author; (*bottom*) MOMA photograph by Sharon Broder. Page 5: (*top*) salt sentinel rendering by Gerd Schmieta; (*bottom*) Vitality card photograph courtesy of the author. Page 6: (*top*) dog collar photograph used by permission of Tagg; (*bottom*) Goodnight Home used with permission of Goodnight Home. Page 7: (*top*) Flower Power photograph used by permission of Flower Power; (*bottom*) August Smart Lock photograph used by permission of August Lock, design by Yves Behar. Page 8: (*top*) Sensoree photograph used by permission of Sensoree; (*bottom*) Red wheel photograph used by permission of Superpedestrian.